FLOODSHOCK

Flood rescue, Scranton, Pennsylvania, 1955.

FLOODSHOCK

THE DROWNING OF PLANET EARTH

Antony Milne

ALAN SUTTON
1986

ALAN SUTTON PUBLISHING
BRUNSWICK ROAD · GLOUCESTER

First published 1986

British Library Cataloguing in Publication Data
Milne, Antony
Floodshock: the drowning of planet
earth.
1. Floods
I. Title
551.48′9   GB1399
ISBN 0-86299-270-2

Jacket picture: Stockphotos Inc.; photograph: Larry Pierce.
Endpapers: (front) 'The Great Deluge' by Sadeler; (back) Canvey Island floods, 1953.

Typesetting and origination by
Alan Sutton Publishing Limited.
Printed and bound in Great Britain.

Then must my earth with her continual tears
Become a deluge, overflow'd and drown'd.

Titus Andronicus

CONTENTS

	Introduction	1
1	The World's Most Famous Flood	5
2	Catastrophe!	15
3	The Big Melt	30
4	Remains of the Legends	40
5	Floodwave!	50
6	Stormshock!	68
7	Earth's Rampaging Rivers	80
8	The Drowning of America	101
9	The Flood Engineers	119
10	Man the Floodmaker: Slashing, Burning and Melting	136
11	Our Teetering Climate	151
	Epilogue	161
	Postscript	165
	Bibliography	167
	Photograph Credits	171
	Index	173

INTRODUCTION

Is Planet Earth becoming a more dangerous place?

The evidence, since the last war, seems to put the answer in the affirmative. Natural catastrophes are occurring with disturbing frequency, with more and more people being killed each year. In the 1970s the annual average death toll from disasters, ranging from earthquakes to floods, was 142,000. In the 1960s it was only 22,500.

One obvious reason for this phenomenon is population growth, especially in the Third World. In poorer countries there are some 3,000 disastrous deaths every year, while in the West there are only about 500. Because of poverty many victims in the underdeveloped nations die after their flimsily built houses collapse around them or are swept away. The inhabitants of poorer countries, by a tragic geographical irony, are more prone to earthquakes and fierce cyclonic storms than those living in the rich northern hemisphere. They are forced to live on dangerous ground or low-lying river deltas. In Rio they live on steep slopes that often wash from under them, and the poor suburbs of Guatemala City are highly susceptible to earthquakes. Often natural disasters simply magnify the political incompetence of the ruling administration, or, as in the case of the floods in Bangladesh in 1970, precipitate appalling fratricidal warfare.

But why a book about floods? Some readers, having experienced or read about droughts in many parts of the world, and having heard much talk about a climatic warming, may have erroneously come to the conclusion that the world is in fact drying out. But frequent droughts are evidence of climatic aberrations. Floods are the interface of droughts, the other side of the same coin. Recent prolonged dry spells in Australia and the far east, for example, have only come to an end with fearsome storms that have inundated many thousands of square miles, and caused widespread death and destruction to people, livestock and arable land.

Of course, climatic change occurs quite naturally over the centuries due to orbital tilting in earth's journey around the sun, and to changes in the amount of solar energy reaching the planet's surface. But it is worth reminding ourselves how watery our misnamed globe actually is, and how big a bearing this fact would have both upon the factors that trigger long-term meteorological disturbances and upon the results of such changes in weather patterns on earth's biosphere. Water and moist gases exist in loose combinations in the earth's crust, and water vapour is even ejected into the air in volcanic eruptions. Water

1

is indispensable to life on earth, and is part of life. Nine-tenths of all earthly creatures, the myriads of marine inhabitants, are found in water. The lower animals and vegetable tissue contain 90 per cent water, and the higher animals, including Man, can contain as much as 70 per cent water. Man's food largely consists of liquid; for example it takes 300 tons of rainfall to grow just one ton of corn.

The world, even with all its parched, drought-blighted regions, is saturated with moisture. R.M. Hare, the geographer, once calculated that in a room 25 feet square, with a ceiling height 15 feet from the ground, there is likely to be some seven to ten pounds of water vapour on a normal English summer's day. In the equatorial zones there would be about twice as much, even in the drier spells.

Eventually all this moisture – from respiration, the trans-evaporation from plants, the dehydration of dying tissue – will re-enter the cycle. It ultimately becomes rain, and returns to the world's oceans, lakes and rivers, as does the evaporation from all that surface water. So a permanent hydrological cycle becomes established.

But this cycle is not akin to some quietly functioning perpetual motion machine, operating faultlessly and predictably. Year after year, century after century, great swirling mists of moisture and wetness have ravaged the earth, while rampaging seas have devoured the world's coastlines.

As a result floods remain in the big league of disasters, if only because the death toll of a serious inundation is much higher than for any other. Floods, in fact, affect more people than any other disaster except droughts. But floods are fast heading for the Number One position. For example, in 1978 the Indian government reckoned that one in every twenty people in that hapless land were vulnerable to flooding. Indeed in that same year some 21.6 million acres of scarce arable land were inundated, while 46,000 villages were flooded out.

According to the US Office for Foreign Disasters Assistance, the number of victims affected by floods jumped from 5.2 million a year in the 1960s to 15.4 million in the 1970s. This 300 per cent leap is much higher than that for droughts, which only experienced a 35 per cent increase. Over 1964–82 floods took the lives of 80,000 people and affected 221 million across the globe. Let us look at a few examples of natural disasters occurring in the 1970s:

Location	Type	Date	Estimated Deaths
Bangladesh	Flood	1970	500,000–800,000
China	Earthquake	1976	242,000
Ethiopia	Drought	1973/4	200,000
Sahel	Drought	1970/3	100,000–150,000

Some idea of the awesome magnitude of the flood problem can be gleaned from a mere snapshot taken since the last war. In a research working paper published jointly by the universities of Toronto and Colorado in 1969 details of a comparative study of natural disasters over the 20 year period from 1947 to 1967 were released. This period was relatively free of serious climatic aberrations. Yet it showed that out of a total of 655 serious events, floods, at 209 (accounting for 173,170 deaths, some 8,600 a year), topped the list of a range of fifteen separate categories. The next most serious types of disaster, which, according to this study, were typhoons and hurricanes, numbered a mere 148.

And yet if one includes all aqualoid disasters – connected with moisture or water – the total of flood-related events rises disturbingly. For example, during the 20 year period thunderstorms accounted for 32 disasters (20,940 deaths), snow storms 27 (3,520), rainstorms 10 (1,100) and tidal waves 5 (3,180). Hence out of 655 individual natural disasters, the total of purely aqualoid events rose to 373, nearly three-quarters of the total.

Strictly speaking, however, one cannot exclude any meteorological excess, since they all tend to contribute to, or make worse, the flooding scenario. The total of these during the 20 year period, including heat waves and cold snaps, amounted to 243, thus raising the total further to 616 – *nearly 90 per cent of the total*. Even that does not end the putative flood connections. There are more than a million earthquakes recorded globally each year (most of them merely seismic tremors). Yet about 100 of them take some 15,000 human lives, many of them because of the resulting tsunamis (or tidal wave).

Now, population growth by itself threatens to bring about more floods. The world's cities are getting larger, and they are pumping more heat and moisture into the atmosphere – and both heat and moisture are the prime instigators of wet and windy weather. In addition the emission of increasing amounts of carbon dioxide into the global skies is bringing about the well-known Greenhouse Effect which prevents much of this additional terrestrial heat from escaping into space. Recently America's Environmental Protection Agency has warned that the Greenhouse Effect, by the middle of the next century, could melt the Antarctic ice packs and raise the ocean levels by several feet or more, to inundate coastal cities around the world.

This is not something that can be dismissed lightly, since scientists have perfected highly accurate techniques for measuring all the variables involved. We know for a fact that the earth's ice packs have grown and shrunk many times in the past – without any additional Greenhouse warming – with remarkable speed. Indeed, if we or our descendents are to be swamped with melting icecap water we may be reliving the traumas of our early partiarchal forebears, as I shall argue in this book. We must not overlook the fact that in both the Judaeo-Christian and the Chinese cultures the creation of the world started with a Great Deluge, the exact nature of which has been the subject of speculation for decades. Perhaps the myriad Flood legends, through the process of mythological distortion, were meant to be warnings.

We must ensure that we do not ignore them.

Chapter One

THE WORLD'S MOST FAMOUS FLOOD

The entire human race possesses extraordinary written legends about a catastrophic flood that engulfed huge areas of the planet earth. The vast amount of ancient documentation in the form of myths, poems and narratives, not forgetting the hundreds of books written on the subject in no less than 72 languages, illustrates the enormous hold that this great inundation has on Man's imagination even today, thousands of years after it is supposed to have taken place. We still, in fact, use the word 'antediluvian' to speak of history before that point. The Bible commences its narrative of earthly history with Noah's flood, and Chinese history also starts from a similar great flood.

A full account of this watery annihilation of apparently all existing civilization can be found in the book of Genesis, from chapters 6 to 9. But the Sumerians repeat almost word for word what is said in Genesis. Occupying Mesopotamia around 5,000 BC, they left behind for posterity their Epic of Gilgamesh in which they also tell of a 'deluge'. The narrative differs only slightly when Noah is replaced with a man called Utnapishtim. Again we read of the building of an ark, and the loading of it with various animal species together with human families and supplies of food. And again there is the landing of the ark upon a remote mountaintop.

The connection of the Great Flood with the dawning of history derives from the central place that the Tigris/Euphrates Valley has in providing the first evidence of cities, temples and a literate civilization. These epic stories from what is now the Middle East probably first reached the Hebrews and Greeks from separate Sumerian and Babylonian flood stories, both of whom incorporated a Deluge legend into their own sagas about the origins of the earth.

Sir James Frazer, the anthropologist, once drew attention to the fact that Noah was the tenth descendent from Adam, while Utnapishtim was the tenth king of Babylon. Indeed, common sense tells us that the Aryan races living in the

The Great Flood legend, with Noah's Ark being tossed on storm-rent seas, as depicted by Sadeler.

Artists tried to faithfully reproduce the Bible's assertion that the Gread Flood was the product of incessant rain, but the more imaginative arists, like Martin, implied that storm surges and tidal waves played a major role.

Levant would absorb the cultural memories of adjacent communities. The well known Armarna letters, discovered during an excavation in Syria, virtually prove that the Israelites only developed a Deluge legend after hearing of similar Babylonian stories.

Where did it all start? The ramifications of the Flood legend may have been largely due to the assiduousness of Ashurbanipal I of Assyria (668–626 BC). He was known as a scholarly and benign ruler, who built a great library containing over 20,000 tablets at Nippur in the Lower Euphrates Valley. These, in the main, were transcriptions of earlier bibliographies describing events well before 10,000 BC. A great many of these tablets related traumatic histories of ecological disasters and inter-communal conflict.

By a stroke of luck, we have direct evidence of what was written on some of the tablets from Ashurbanipal's library. A broken tablet was discovered in 1872 by a young official from the British Museum. It recorded, among other things, an account of a ship alighting on Mount Nasir, and of the sending forth of a dove. There was much reference to a hero, a priestly king, who was an esteemed and worthy man not unlike Noah. But God warned the hero, and there is a story of a great ship, and of feverish planning, but of ultimate drowning. This sliver of clay was the shattered remnants of the world's first Doomsday book.

The Strange Case of the Global Noah

So was Noah's flood merely parable, myth, legend? Many anthropologists believe not, because many flood legends describing cataclysmic inundations are to be found across the world. They have been discovered in India, China, The East Indies, Polynesia and most significantly in the Americas. What is puzzling is that they emanate from great continental land masses where the inhabitants have never seen the sea, nor great rivers or lakes.

Under various aliases Noah hauntingly reappears from every corner of the globe, with his poetic message of deep, stormy waters and doves and ravens. Indeed, the flood researcher might be excused for assuming that the inundation first occurred in the western hemisphere, to be later transmitted to the east. From the Canadian North-West south of the Arctic Circle, down through North and Central America and into the southern continents, virtually every tribe and ethnic group has a legend referring to a worldwide flood.

The Spaniards, it seems, were overwhelmed with Deluge traditions when they arrived in the Americas. To the Catholic invaders some of the legends, so closely resembling the Bible story and yet clearly of alien origin, seemed sacrilegious. The reason for the floods in all these Amerindian fables is invariably the same: mankind became wicked, and God decided to destroy the species, but at the same time to save one good couple or family and allow them to start afresh and build a new and better civilization. The Jibaro Indians of the Upper Amazon tell that 'a great cloud of rain fell from heaven, like a sheet of water, and caused the death of every living thing on earth'. This, in its most explicit form, was what the Incas told the Spanish in Peru, even saying that the rains lasted for sixty days and nights.

So incensed were the Conquistadores with these blasphemous parodies of the Bible story that they ordered many of them to be burnt. This was the fate of the *Popul Vuh*, a Quiche Maya chronicle written in hieroglyphics. Copies inevitably survived, and it was soon translated into Spanish by an obliging monk. It read, in part, 'The waters were agitated by the will of the Heart of Heaven (called Hurakan), and a great inundation came upon the heads of these creatures . . . the face of the earth was obscured, and a heavy darkening rain commenced, rain by day and rain by night.'

Another Indian myth that survived desecration was the Hopi description of a country where great cities evolved in which many different crafts were undertaken. But when the people became corrupt and warlike the Great Flood inevitably brought things to an end: 'Waves higher than mountains rolled in on the land, and continents broke asunder and sank beneath the seas.' The Chibcha Indians, in Colombia, have a similar tale that describes how the deluge was initiated by a god called Chibchacun of whom a higher god had punished by forcing the Noachian hero – Bochica – to carry the whole earth on his back.

The ancient Aztec picture-written documents speak of a Mexican Noah called Tezpi. He, too, had saved himself and his wife in a boat made of cypress wood. The Zapotecs, however, have traditions more strikingly in conformity with the Genesis story. They tell how Tezpi embarked in a spacious vessel with his family and several animals, and food, in order to continue the human race after the disaster was over. When the great god Tezxatlipoca decreed that the waters should at last subside, Tezpi released a vulture which failed to return. Tezpi then sent out other fowl, and only the humming-bird came back with a leafy twig in its beak. This final proof of renewed vegetation led Tezpi to abandon his raft on the mountain of Colhuacan.

The preoccupation with mountains uncannily strengthens the Biblical record. In almost every case the famous Noah's Ark came to rest not only 'upon the mountains of Ararat' but upon rocky edifices everywhere, even man-made ones. The Aztecs even claim that the pyramid of Cholua, in Mexico, is an ancient refuge against future floods. The Aztecs believed in World Ages, whereby civilizations rose but were destined to plummet to universal destruction. Their *Codex Chimal-popoca* relates: 'The sky drew near to the earth and in the space of a day all was drowned. The mountains themselves were covered by water. It is said that the rocks we can see today rolled about over all the land dragged by waves of boiling lava, and that there suddenly arose mountains the colour of fire.'

The flood was deep and catastrophic, and covered every bit of dry land except the mountain tops, representing a last gasp chance of survival. Today we cannot conceive of such a flood, but in mythology the jutting peaks represented the gods' hand of rescue: all was not lost. The most fortunate of the earth's doomed sinners actually lived in and around a mountain – such as the ancient peoples of Ancasmarca, in the Andes. In this fable a shepherd and his family assembled all the food and livestock they could, and brought them to the top of the mountain to avoid a forewarned inundation. The Ancasmarcans were indeed blessed: as the teeming waters rose ever higher, the mountain peak rose up in unison, thus sparing the occupants from drowning. The shepherd and children then repopulated the land of Ancasmarca.

Indeed, the entire Andes region, it would seem, was the best place to be at the time of the Great Flood. In his *The Discovery and Conquest of Peru*, published in 1968, Augustin de Zarate, a treasury official based in Lima, said that the Indians believed that their ancestors, warned of a massive flooding, took refuge in spacious mountain caves. Survivors of this cataclysm from many races returned to the Andes, and by tradition this area became 'Tsa huantin-Suyu' – 'the common gathering place of all nations'. This was the title later appropriated by the Incas for their 'Golden Empire' of Peru.

Occasionally the colourful Amerindian sagas revert to plausible creation myths, with an acknowledgement of the formation of oceans from aeons of torrential rain, after which dry land and a breathable atmosphere enables human life to evolve. Some scholars, reflecting a division of opinion as to the putative cause of the Flood, use the word 'deluge' to refer to years of incessant rain in line with this geological perspective. Others look to more catastrophic events, as we shall examine in the next chapter, that could have caused giant tidal waves and massive upheavals in the earth's crust.

Indeed, some North American Indians possess flood legends that are more recognisably catastrophic. Mexican traditions speak of earthquakes, tidal waves, erupting volcanoes, cities swept into the sea, and millions of people being destroyed, the lucky survivors being those who had fled to outlying caves. The Guarani Indians used to tell stories, repeated by white settlers, about Tamandere who, when the rains became torrential, stayed in the valley instead of fleeing to the mountains with the rest of the tribe. Tamandere clambered up a palm tree, which became uprooted, and he and his wife rode on it to the top of a mountain to join the others. After they heard the beating wings of a bird they knew the tide was abating, and then started to repopulate the earth. The Sac and Fox tribe of Iowa and Oklahoma retell a story about rain with drops appropriately 'as big as a wigwam'. (In the Book of Revelation, where there is a reminder of the Great Flood, we similarly read that 'a great hail fell out of heaven; every stone about the weight of a talent'.)

The Hawaiian Islanders have long come to terms with their own traditions of earthly catastrophe. There is a familiar ring to the story: the first man became so wicked that his god, Kane, decided to eliminate him and tear down the earth on which he lived. Earthquakes, cosmic bombardment, volcanic eruptions were apparently meted out regularly, since each new man was as evil as his predecessor. Kane, however, soon became weary of building and rebuilding parched and shattered universes. So he tried a different approach. He decided to allow one clearly righteous man, Nu-u (remarkably like Noah), together with his family, to erect a 'great canoe' with a house on it. Kane then advised Nu-u to take his wife, Lili-nu-u, and their children and all the animals he needed, to sit out the flood.

When the rains fell the waters rose and spilled into the oceans which seemed to merge and coalesce. The giant house-canoe drifted for days, while below the whole of mankind was obliterated. Finally, Kane signalled that the worst was over, and the torrential downpour eased. Glimmers of sunlight could be seen, causing a magnificent rainbow which Hawaiian legend has it was Kane's gesture towards atonement and forgiveness for the terrible thing he had brought about. Then Nu-u and his wife were told to repopulate the earth.

The Chinese also have earth-shattering Flood sagas. They declare that the planet was not only totally submerged in water, but that 'pillars of heaven' were torn asunder. A great shaking affected the land, and it fell to pieces. At the same time strange things were happening in the sky as the northern heavens became sunken whilst the sun and moon changed their motions.

The Chinese hero escaped the destruction, as did his three sons, daughters and wife. Once again the ubiquitous Noah resurfaces in Chinese garb. All of the Chinese Flood stories contend that the people of mainland China are direct descendants of an ancestor of great antiquity called Nu-wah, famous for having survived a great drowning. It is intriguing to note that the word for 'ship', as printed in contemporary Chinese books, is the ancient character made up of the picture of 'boat' and 'eight mouths', showing that the first ship was a vessel carrying eight persons.

Arks, mountains, godly retribution; time and again the world's extraordinary Flood legends repeat common themes, with hardly any pronounced differences that would mark them out as not belonging to the Noahchim genre. The Middle East, the Orient, America all have them. They all talk not so much in terms of the world coming to an end, but rather in terms of a New Beginning after a great purging or cleansing. More than thirty Flood fables have been discovered over the years in the Orient, all with identical messages. The Battaks of Sumatra speak in laudable terms about the Creator, called Debata, who repented of his solution to finish off all mankind, and allowed the waters to subside. Whereas the Sumatran legends, dispensing with the Ark but keeping the mountaintop hideouts, are similar to those of the Aztecs, the Hindu legends are strikingly similar to those of the Hebrews and Chaldeans. Any scholar who has studied the Indian manuscripts such as the Rig-Veda and the Upangas is quite familiar with them.

The anthropologist will give a simple explanation: they have all originated from Ashurbanipal's library via a process known as 'cultural diffusionism'. In other words, a very good story, one worth telling to one's offspring to warn of the perils of bad behaviour, has somehow spread itself around the world. Important clues can be detected from the various narratives. The closer ethnologically a tribe is to its Biblical ancestors, the more the small print of Genesis is adhered to. The purpose of the ark, after all, was to house a great number of animal species. In many legends the emphasis is on human survival and procreation. But the Iranian legend relates how Yima, the father of the human race, was warned by Ahuramazda, the Almighty, of the flood. He ordered Yima to build a *vara*, or vessel, to contain the seeds, or sperm, of varieties of domesticated and wild beasts so that they could be regenerated later.

The rest of the world borrowed the legends and embellished them, either overriding their stranger features or emphasizing them. In the Welsh legends two lone survivors, Dwyfan and Dwyfach, rescued themselves in a craft 'without

A great many Flood legends repeat an important feature in the Genesis narrative: a bird was released from the Ark to return with proof of new vegetation, hinting that the flood was abating. Anthropologists believe the Ark legends spread themselves around the world via 'cultural diffusionism'.

rigging' (i.e. an unusual-looking vessel). The Scandinavian races could only make sense of the Flood fables after a good deal of adaptation. Three sons of Borr allegedly slayed the father of the 'ice giants', a god named Hrimthursar. There was so much blood about that the entire race of giants was drowned with the exception of Bergelmir, who saved himself and his family in a boat, and reproduced the race.

One great difficulty for the ethnologist is to disentangle the primitive narrative of observable phenomena from mythology, or to discern the grains of truth in an otherwise grotesquely embellished saga. The coastal area of Chile, near Conception, has a lengthy flooding history, which has given rise to just such 'Great Flood' tales among the native Araucanian Indians. Most of these floods were really *tsunamis*, i.e. gigantic sea waves precipitated by earthquake activity. In 1751 Conception was completely destroyed in this way, and a series of like disasters can be traced back to the sixteenth century.

Archaeological investigations, too, suggest that the Chinese floods were an exaggerated account of a local flood or violent storm. Certainly there was flooding on a large scale. But then China has had a long and wretched history of extremely severe inundations largely stemming from river overflows. Indeed, Chinese legends hint at this when the catastrophic overspill of the 'great rivers' was said to be halted by the eventual swelling of the sea.

It has been suggested that modern Christian missionaries have been the carriers of the Noahchim stories which have been grafted on to vaguer ancient traditions. Some anthropologists dispute this by pointing to the fact that the missionaries, as we saw when the Spanish arrived in Latin America, were

11

themselves surprised at the similarities of the legends even before they began teaching. The Navajaro Indians, for example, believed the Grand Canyon to be the product of a deluge long before the first contact with Europeans.

Curiously Africa, with a long history of missionary proselytizing, doesn't have a Flood legend. And some cultures are aware of the Great Deluge legends, but claim that their communities were not affected. When the Egyptians related the Flood sagas, they were at pains to point out that they were spared. As the Egyptians were an important people amongst those of the Aryan race, and lived adjacent to the apparently worst affected areas, this is an intriguing and baffling anomaly. Theirs was the lone voice of moderation in a world of shrill doomwatchers. Their rebuke of the Hellenes is of great interest to geographers. The Hellenes were accused of being childish in attaching so much importance to the Flood, saying that there had been several other local catastrophes in the past.

The Fabled Continent of Atlantis

Any discussion of global Flood legends cannot fail to include the extraordinary prolific writings on the lost continent of Atlantis, often thought to lie beneath the present Atlantic Ocean. It was the Greek philosopher Plato, alive some 400 years before Christ, who first brought Atlantis to the attention of the civilized world. He wrote the famous *Critias*, which was a narrative account of a conversation held between Solon, the eminent Greek lawyer, and Plato himself.

Solon, it seems, had earlier gone to Egypt and had met some influential priests. They had told him that a land as big as Libya and Asia (presumably they meant Asia Minor) joined together once existed, and indeed flourished, and that it was 'not far from the Pillars of Hercules' (the Straits of Gibraltar). Solon was particularly interested in this story, as the Egyptian priests had intimated that Atlantis was partly populated, or at least occupied, by Greeks. He was alarmed to learn from the priests that there was a violent earthquake, and that 'in a single day and night of rain all your (Greek) warlike men in a body sunk into the earth, and the island of Atlantis in like manner disappeared. . . . '

In his *Critias* Plato goes into some detail about the wealth and splendour of the luxuriant vegetation and the splendid climate. There was an indigenous range of wild animals, including elephants and equine creatures. In another work Plato gives some approximate dimensions of this marvellous but vanished land. He said it was oblong, with a steep coastline. The capital was architecturally interesting, with various concentric rings of land, and had an important harbour.

The connection between the Great Flood legends and Atlantis is more closely linked with the traditions of the Aztecs. By all accounts the Aztecs told the Spanish conquistadores that their people were descended from a people called Az, and came from Aztlán, a sunken land in the east. It has been pointed out by linguists that the Aztec word for water, *atl*, also means the same in Berber. This is the language of an Aryan race inhabiting the Atlas Mountains of North Africa.

The Mayas of Central America had a tradition of a white god called

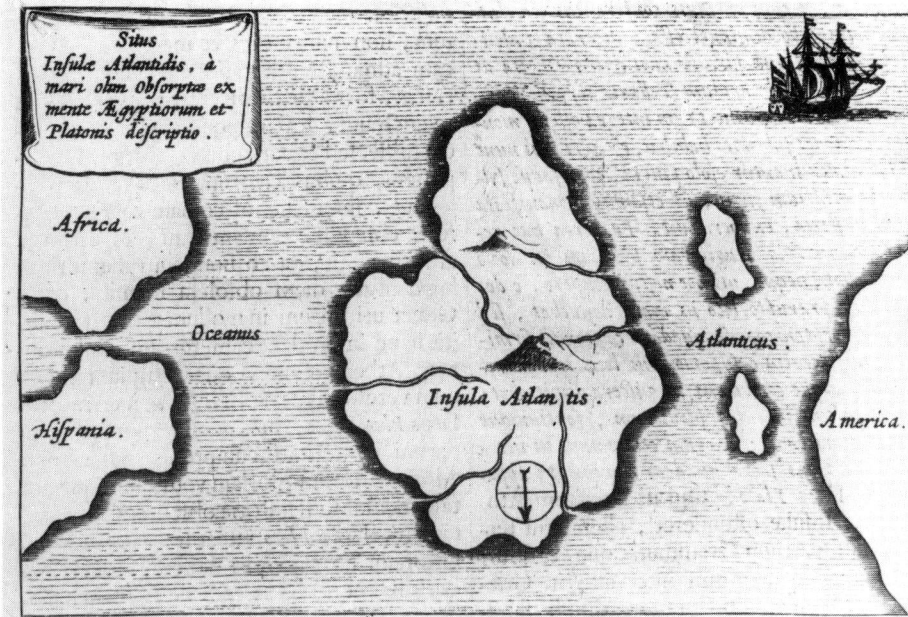

The 'lost' island of Atlantis, as its name implies, has been traditionally sited in the Atlantic Ocean. Curiously, this 400-year-old map has been drawn upside down.

Kukulkăn, who bestowed a benign civilization upon the New World. More importantly, the Mayas have a strong tribal memory of Aztlán, as did the Venezuelan 'white' Indians who were, it is believed, eliminated by the Spanish invaders. Many American Indians worship Quetzalcoatl, said to be an old white man with a long beard, who arrived in the Valley of Mexico from the ocean to embark on a humanising mission.

The Atlantis myth has similar echoes elsewhere. The ancient Aryan saga, the Mahabharata, says that about 60 million people 'in great cities' were killed in one dreadful night. More specifically the Troano Manuscript of the Mayas says most of these lands were to the west, placing them in the Pacific where some writers believe the lost island of Mu was, located somewhere in the area now occupied by the South Seas. 'Twice upheaved', reads the manuscript, 'they broke into ten pieces and sank, together with millions of inhabitants.'

The Chinese similarly cherish traditions of a huge island called Maligasima, destroyed yet again owing to the evil of its giants. Again a King Peirium escaped, like Noah, and his offspring peopled China with Divine dynasties. The Hindu Puranas offer a vivid description of wars on continents and islands situated beyond West Africa in the Atlantic Ocean.

Is that where Atlantis lies today? Thousands of books and articles have discussed this point interminably. It is at this stage that legend must be disentangled from other apparently factual accounts of similar-sounding populated islands of ancient history. For instance, the Carthaginians and their

13

predecessors, the Phoenecians, were said to have visited a great island called Antilla, which often appeared on medieval maps and probably represented the present-day Azores. There is a wide consensus of opinion among Atlanteans that the Azores are in fact the remaining tops of the mountains of the sunken Atlantis. Significantly Atlantis was reputed to have hot springs, which today can be found in the Azores. Some cartographers, however, deny that Antilla and Atlantis were one and the same, as the Azores had already been discovered by the time Columbus made his first voyage, and took with him the Benicasa Map, on which Antilla was placed in the middle of the ocean.

The combined Great Flood/Atlantis legends are highly allegorical. From all the sagas it appears that the human species endured a false start. Atlantis might have once been a rudimentary island civilization of Toltecs, but it grew in the collective imagination into a populous and mighty nation. Its splendid and godlike people became decadent, but not before traversing the western world from the Atlantic coast of the Americas to Asia Minor, founding the Aryan race and spreading wisdom. But Atlantis and the other elaborate civilizations were destroyed in a global cataclysm and had to start anew, relearning virtually everything.

Hence most observers view the legends as a warning to the rest of mankind. The omens are clear: your entire world could be lost forever by malign or uncaring behaviour. The dominant theme is an historic one of ethnic struggle, defeat in war, but ultimate triumph. Underlying it all, however, and superimposed upon human tragedy is a convulsion of nature, so terrible and irresistible that it can never be allowed to be forgotten. In each legend there is talk of destruction and recovery, of some kind of cosmic upheaval. A divine being is invariably associated with the awesome event, and there is a sense of profound geophysical disturbance. Two startling truths are revealed in the Flood legends, one cruel and the other inspirational: one is that corrupt deeds bring about destruction from the gods (or the elements), the other that there will always be survivors. Some tribes, like the Egyptians, dimly recognized that this was a recurrent feature of earthly life, something to be borne with bravery and solicitude.

How much is fact and how much fiction depends on the feelings of the observer. To Ignatious Donelly, a nineteenth-century US Congressman who spent most of his spare time researching Flood legends in the Library of Congress, the truth of the floods was painfully obvious. At Athens and Heirapolis, he pointed out, pilgrims came to pour offerings into the fissures of the earth to appease an earthquake god. 'It is too much to ask of us', he wrote, 'to believe that Biblical history, Chaldean, Iranian and Greek legends signify nothing, and that even national festivities were based upon myth.'

The next three chapters of this book will be written as if the floods really did happen.

Chapter Two

CATASTROPHE!

Ever since Darwin geologists have not been too happy about investigating the Deluge legends. Many scholars, scientists and others of a rational disposition are similarly diffident about researching Noah's Flood, not least because few are absolutely certain whether the Genesis story is supposed to be a form of historical documentation or some obscure piece of theology. And the Atlantis legends of land masses sinking violently into the sea have thrown up some other uncomfortable rubrics. The truth is that too many intellectual questions have arisen concerning the relationship of earth to other matter in the Cosmos. We still have no clear-cut knowledge of the early geophysics of the earth, so interpreting past earthly phenomena still entails a lot of educated guesswork.

But if, for the sake of argument, the Bible and other legendary narrative *are* history, what are we to make of them given our present state of knowledge? How credible are some of the theories put forward to explain the Great Flood?

Prior to Darwin and Lyell, geography and geology were infant sciences, usually left to the explorer and excavator. Since the Darwinian revolution there has been a profound schism of geological opinion. Some believe the crust of the earth was formed by an excruciatingly slow process of evolution, with each terrestrial feature – the rocks, valleys and rivers – all changing at the same, slow 'uniform' rate. But others, now in the minority, still believe the earth's birth and maturation has been painful and traumatic, wrought by violent and catastrophic upheavals and celestial bombardments.

The debate has lost a lot of its impetus because the two views have tended to converge as a result of considerable scientific advances made in recent years. But some thirty years ago the debate was bitter and controversial. Because of Genesis the 'catastrophists' were vaguely associated with the validation of religious myths and the wrath of God. We must never forget the extraordinary furore that accompanied the publication in the 1950s of Emmanuel Velikovsky's *Worlds in Collision*. This rumpus was engendered by the frustration of knowing his explanations for earth's geological puzzles were wrong, but not having sufficient knowledge of geophysical history with which to contradict him. We will refer to Velikovsky's theories a little later.

The Uniformitarian perspective dates only from the late eighteenth century, whereas the early scientists were immersed in the heritage of Catastrophism bequeathed them by anient Greek philosophy. Metaphysical discourse was for centuries coloured by the Greek cosmic outlook of the Ionian school, and by

Plato's theories that there were, and would be again, periodic annihilations of the earth by fire and flood. The Egyptian priests in Plato's story pointed out that Phaethon harnessed his father's (the sun's) chariot, but lost control of it. This mythical version of what scholars consider to be factual events implied that the course of the heavenly bodies change their positions. And it is this alarming phenomenon that brings about earthly doom.

The sun's reversal in the sky was a prominent theme amongst great and ancient poets. It is vividly described in Ovid's *Tristia*, ii, 391, and *Ars Amatoria*, i, 32, and *Martial*, iii, 45, confirming at least that the myths of ancient times were known to the Greeks and Romans. 'Truth lies in the fact that the heavenly bodies moving around the world deviate from their courses, and after a long time the earth is scorched and consumed in huge fires,' announced Plato.

The Hindus, Mayas, Greeks and even the Irish, we learn, recall four World Ages preceding our own. All of them were destroyed before slowly being rebuilt. Egyptian priests told Herodotus that 11,000 years before the axis of the earth became displaced, 'the sun had removed from his proper course four times and had risen where he now setteth and set where he now riseth'. The Eskimo legend is even more specific, claiming that the earth tilted violently before the Flood. The oldest Chinese records speak of a time when the sky suddenly began to fall northward. The sun, moon and planets left their usual courses. There are records of other astronomical deviations. The Harris Papyrus refers to a cosmic cataclysm that overturned the earth, as does the Hermitage Papyrus of Leningrad and the Ipuwer Papyrus.

This doom-laden legendary heritage was a constant deadweight upon the struggling empiricism of the early geologists who kept finding curious fossil remains and confusing objects dug up from the ground by farmers and miners working below the surface. The Swiss naturalist Charles Bonnet (1720–93), held that the fossils were the remnants of extinct species, and believed they had been destroyed by planetary disasters. The fossils, said the French paleontologist Baron Georges Cuvier (1769–1823), were the product of four catastrophes, the last being the Flood. Cuvier believed the sea had overwhelmed the land, drowning herds of mammoths in the process. Studying gypsum formations around Paris, he concluded that marine deposits were stratified with others showing terrestrial and freshwater forms.

The fossils, of course, have always been a conundrum. To Herodotus and other Greeks the petrified remains of marine creatures were proof enough that the world must have been under water. But to many Greeks not of the Ionian cosmic school the fossils, looking like little stone doodles, were practical jokes on the part of nature. This idea survived until the thirteenth century, when the Biblical deluge became the main causative agent in the creation of the earth.

Even today the Rockies provide evidence favouring the catastrophist with their thousands of ripple marks, trilobites and other fossils preserved without the slightest trace of disintegration, hinting that they were thrown down abruptly and suddenly. The German philosopher Gottfried Leibniz (1646–1716) believed the fossils to be the bones of sea monsters hurled onto the land by the sea. In one of the early formations of a catastrophe theory Leibniz saw the need for

Gottfried Wilhelm Leibnitz, 1646–1716, German mathematician, philosopher and historian. Leibnitz believed that the Biblical flood was the product of catastrophic causes.

Aristocles Plato, 427–347 BC, Greek philosopher. Plato believed the earth was frequently destroyed by fire and floods.

some other mechanism than the Biblical forty days and nights of rain. He suggested a world of huge water-filled hollows which had suffered enormous crushing forces to flood the surface of the land.

Following Leibniz, John Woodward of England's exclusive Royal Society travelled all over the country in search of unusual stratifications and mineral formations, even turning up at collieries and other subsurface works, chatting to miners and making copious notes. Woodward, in his 1695 essay *Towards a Natural History of the Earth*, came inexorably to the conclusion that some kind of traumatic geological event – probably the Biblical flood – was responsible for the incongruities in rock and earth formations. This event even caused the great mountain chains to come into being, as they all bore telltale signs of ocean life that existed thousands of years before.

However, the age of enquiry had already flickered into life, and geological and tectonic questions needed urgent and plausible answers. Under the impetus of the European Enlightenment there was an incipient but vain struggle away from the doctrines of the church. In spite of themselves, scholars became drawn to the cataclysm of Noah's story. In any event they were not the dispassionate investigators that we would today take for granted. Many remained catastrophists merely because they were steeped in Bible lore. For example, the English geologist William Buckland was also a theologian, and he described a world created by a series of massive upheavals. In 1823 he confirmed his agreement with the Biblical interpretation of earth history. The Flood, he wrote, swept away all the quadrupeds, tore up the solid strata of the earth, and 'reduced the surface to a state of ruin'. Everything was explained, in Buckland's eyes, as 'the direct agency of Creative interference'. William Whewell, a colleague of Buckland, also wrote, with deadly earnestness, of 'creative powers'.

Blazing Comets and Frozen Ice-Moons

Velikovsky, however, is probably the greatest exponent of the colliding comet theory (Hans Bellamy in 1945 advanced a similar controversial comet or asteroid theory). A Russian Jew born in 1895, he studied medicine at the University of Moscow. He later moved to Palestine and became interested in psychoanalysis in which he gained his doctorate. It was when he was researching a medical thesis on a Biblical theme that he noticed the extraordinary number of references to celestial phenomena. He came to the conclusion that a massive comet must have appeared in the skies in the third millennium BC, and he connected it with the Biblical Exodus. After roaring past the earth, causing the planet to tilt and bringing about weird atmospheric disturbances, the north polar ice fields were shifted further north. The gravitational attraction of the comet, which shortly afterwards went into orbit as the planet Venus, generated enormous tidal forces that hurled whole oceans over land masses. Earthquakes and volcanic eruptions ensued, and tremendous winds were unleashed.

The point about Velikovsky's remarkable books is that they are based on the worldwide Flood legends discussed in Chapter One. He amasses a great deal of carefully documented evidence to support his theory. He cites the darkness, rains of stones, the deluges and conflagrations found in the ancient fables as empirical documentation of real and terrifying events. He draws attention to the relative newness of the planet Venus, arguing that in many ancient astronomical charts it is not shown at all. He discusses the change in calender reckoning in both the western and eastern hemispheres: the Mayan year and the Babylonian calendar both start on a date approximating to February 26th.

Velikovsky asserted that it was not always the case that the planets in the solar systems orbited in their present positions, as collisions between them occurred, having the effect of bouncing them from one orbit into another. And he pointed to two important cosmic events in earth's recent history. The first occurred some 3,500 years ago, and the second 2,600 years ago. During the first episode, he theorized, earth entered the outer edges of the careening object's trailing dust and gases. Parts of the earth's surfaces were heated to such a degree that it became molten, and great streams of lava welled out. A thick liquid, naptha, fell torrentially and formed the world's present oil deposits. The sea boiled and evaporated while entire mountain ranges collapsed, and continents sank in cataclysmic floods. The hydrocarbons in the comet's massive tail drenched the earth with petrol as the naptha slowly distilled on contact with the atmosphere, or was vaporized by the incessant electrical discharges that turned it into burning sulphurous lumps of matter.

The later phenomenon, which was apparently recorded in annals from Iceland to India, occurred when the comet returned and careered past Mars which was shifted nearer to the earth and nearly collided with it. This second celestial event also succeeded in tilting the earth and reversing the rotation of the axis, as confirmed in the celestial charts, sundials and calendars of ancient races.

So at least twice in Subboreal time – from about 2,000 BC to about 8,000 BC – havoc was wreaked on the globe. Valleys were torn out, lakes overturned, forests replaced with bogs, and human tribes were decimated. In his later book, *Earth in Upheaval*, Velikovsky writes: 'Tidal waves traversed continents, moving by inertia when the daily rotation of the earth was disturbed. . . . These tidal waves, augmented by others produced by the extraneous fields of force generated by boulders, distributed marine sediment over the land . . . and floods (were) caused by the melting ice cover . . .'

We must remind ourselves that Velikovsky's theories are totally at odds with current understanding of planetary chemistry and physics. Even in 1940 there were important scientific findings that could explain the high temperature of Venus, the main 'proof' that Velikovsky offered to explain why the planet was quite recently in a molten state.

Carl Sagan, in his *Broca's Brain*, has dealt at length with all the astrophysical errors in Velikovsky's books. He says that for Jupiter to have ejected a 'comet' the size of Venus into orbit the amount of kinetic energy required would be equivalent to all the energy radiated by the sun in a whole year. Velikovsky also fails to explain, if the earth's rotation was braked to a halt (let alone reversed), how it managed to rotate again from rest without outside help. And if Venus consists of hydrocarbon gases, why does Jupiter – its mother body – consist primarily of hydrogen and helium? And so on.

Immanuel Velikovsky, author of *Worlds in Collision*. Velikovsky believed that Venus was once a giant 'comet'. Before settling into orbit the comet caused devastating floods around the world.

It seems that Velikovsky, with his astronomical catastrophism maturing in the 1930s and 1940s, was still under the magical spell of 'cometomania' that has affected otherwise rational men for centuries. Comets have always been harbingers of floods and destruction, and even held responsible for wars, revolutions, massacres and much else besides (the science writer Joseph Goodavage, in his book *The Comet Kahoutek*, is quite prepared to go into print saying as much).

21

Today we now realise the error of attributing catastrophic powers to a comet, a luminous celestial body containing – by interplanetary standards – a tiny nucleus. A true comet is enveloped with a gaseous haze or coma that forms the tail, that bright searing streak that observers can see from the ground. This tail is so tenuous that stars glimmer through it, as it is largely comprised of small particles of ice bound together with an extremely thin gas.

But this has not stopped educated laymen from making the same mistake as William Whiston, a clergyman and one-time professor of mathematics at Cambridge, who published a doomsday book in 1696 in which he confused a comet with a planet. As related in Patrick Moore's *Countdown*, Whiston's comet bypassed the earth in biblical times to create huge tides and landslips, broke down mountains, and flooded the earth up to six miles deep. Both the nucleus and the tail of the comet are supposed to have struck the earth and discharged forty days and forty nights-worth of rain.

But Whiston in turn may have been heavily influenced by Edmund Halley, one of the leading astronomers of his generation. For in 1682, a portentous year, Halley's comet appeared in the heavens. It was immediately clear to him that the earth must have been frequently bombarded throughout geologic time, and many scientists thus had their God-inspired catastrophism reinforced. Later the Marquis de Laplace (1749–1827), a genius in mathematical astronomy, believed that a close encounter with a comet would cause the seas 'to abandon their ancient positions' and rush towards what he called a new equator, with much of the world being drowned in the process. Comets were virtually the same as planetesimals in the infant astronomical world of Laplace. He insisted that the motions of the earth were not unalterable, being subject to unpredictable forces, including heavenly collisions. In his *Exposition du Système du Monde* he used two pages to argue that mankind should learn to accept potential cosmic dangers. He said '. . . entire species would be annihilated; all monuments of human industry overthrown; a great portion of the human race and animals would be drowned in the universal deluge'.

A more up-to-date mentor for Velikovsky would have been the Viennese mining engineer Hans Hoerbiger (1860–1931). His *Welt Eis Lehre*, or 'World-Ice Theory', postulated in 1913 that in its long history earth must have had at least six successive moons, each drawn into orbit by earth's gravitational force. Each eventually spiralled closer to the surface, pulling the seas into an immense girdle tide around the equator. Every seventy million years or so a moon would plummet down on to our planet; the last one caused the extinction of the dinosaurs. During the 1930s Hoerbiger's theories were elevated to the status of a cult, and the Nazi leaders – many of them inspired by occultist and mystical teachings – promoted it.

Hoerbiger theorized that space must be permeated with hydrogen and water-vapour. In this, of course, he was only partly correct. The gases in the universe in fact consist of 90 per cent hydrogen and 9 per cent helium, with a residue of 1 per cent of other gases, including oxygen. He made the error of assuming that the moon and other planets had a covering of ice some 150 miles deep as a result of the deep cold of space that would freeze the vapour solid. The

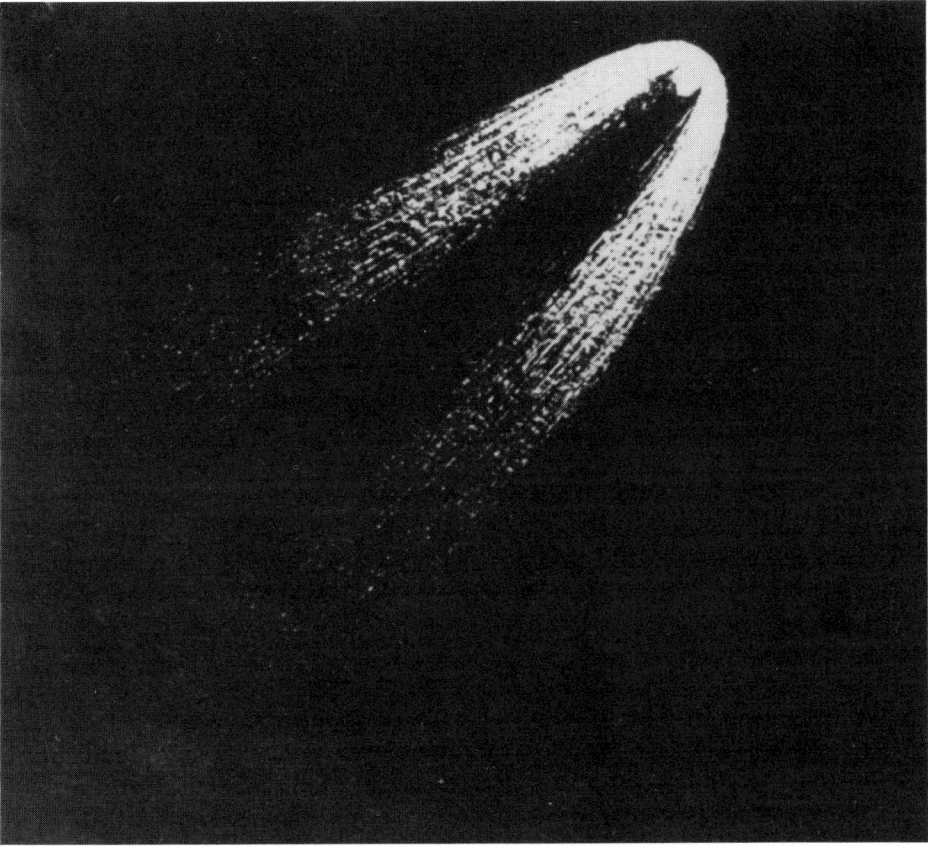

Edmund Halley himself believed that his comet, and many others like it, must have been responsible for worldwide floods, earthquakes and other forms of terrestrial doom.

present moon was captured about 12,000 years ago, and is also doomed to crash to earth. In a previously captured moon, the earth's temperature dropped dramatically by the approach of so large a chunk of ice, until finally the ice-moon was shattered by the powerful gravitational forces. Ice hurtled down upon the land, followed by huge amounts of meteoric debris. The earth suddenly became more spherical in shape, and violent earthquakes and floods followed as the equatorial waters flowed back to higher latitudes, finally plunging Atlantis beneath the waves.

Hans Bellamy, a devotee of Hoerbiger's, wrote not of a comet or a moon, but of a 'close former satellite' which passed by the earth. The gravitation of this celestial body caused the waters of the globe to rush down to the tropical zones, the so-called 'girdle tide'. But the planet was still in an ice age, and the warming of the tropical waters melted the ice sheets as they reached higher latitudes. The Eskimos of those days, he says, 'noted with surprise two things and preserved the experience in their myths: the waters of the deluge were "hot" and they were "salt"'.

23

Edmund Halley, 1656–1742, English astronomer. Halley's comet inspired many doomsday writers to attribute catastrophic powers to all comets.

The Amazing Pole Shift Theories

Velikovsky's key description of the geophysical disturbances, we must remember, was the shifting of the earth's axis. This is a remarkably popular theme amongst Doomsday book writers. Most of the theories about a tilting axis are connected with the ice caps. In Velikovsky's theories the ice packs were thrust further north by the force of the meteoritic impact. But in other theories it is the accumulating *weight* of the ice sheets during the last ice age that brings about the tilt. One speculation is that before disaster struck the North Pole was located somewhere near the longitude and latitude of Iceland, probably at the south end of Greenland.

The assumption is that if the axis did in fact shift, violent earthquakes would rend the earth. Air and water would continue to move through inertia, thus causing the seas to rush over the continents. 'Lakes', wrote Velikovsky, 'would be tilted and emptied, rivers would change their beds, large land areas with all their inhabitants would slip under the sea.'

The late Hugh Auchincloss Brown was an amateur scientific investigator who spent a lifetime formulating 'pole shift' theories. The earth, argued Brown, already has an imbalance in its rotation, the so-called Chandler Wobble. But Brown predicted that the earth would become 'top heavy' when the ice grew more than three miles deep, and would start to wobble like a spinning top as soon as the ice on the south pole became excessively concentrated. Then, when the axial poles moved away from the poles of spin, the earth would actually flip over in space and start a new spin-axis with altered polar locations.

The mechanics of this extraordinary event, as described by John White's *Pole Shift*, are dramatic, even if highly fanciful. The equatorial bulge slips over itself like a wave in the ocean travelling at 100 miles an hour, while the rotation of the axis remains 'upright' because of the gyroscopic stabilization of the earth's spin. Says White: 'An observer standing on earth between the pole and equator along the line of careen would see the bulge heading towards him as a ground wave some thirteen miles high!' After the flip has ended, the ice caps 'end up near the equator', and the presumed rapid melting of them resulted in the Great Flood.

Modern geophysicists suspect that such a 'pole shift' would destroy the earth. Even so, Auchincloss Brown and other fellow travellers assert that this polar flip has happened frequently – every 7,000 years or so – in earth's history. It not only caused the death of the dinosaurs some 65 million years ago but at other times much of the natural ecosystem.

Fears concerning the ice packs at the Antarctic are common in the catastrophe legends. Charles Hapgood, a science professor at Keene College, New Hampshire, published, in 1958, his *Earth's Shifting Crust*. In this he argued that when the poles in the past changed their positions, some continents were moved towards them and others away. There were dramatic changes in climate, together with flooding and ecological destruction. This time, however, Hapgood convincingly blends the notion of *magnetic* pole reversal with the concept of axial spin. Polar flips of the magnetic kind, scientifically speaking, are quite plausible.

Geophysicists at Cambridge, for example, have discovered that at intervals in the past the earth's polar magnetic fields have changed places. Before the reverse takes place, however, the earth's protective shield attenuates, allowing a lethal penetration of cosmic rays into the atmosphere. The reversal can take place at intervals of between half and one million years, the last possibly occurring 700,000 years ago. This was accompanied by the death of a great number of small marine creatures, and possibly many thousands of animals.

But Hapgood went further than this. He believed that one axis could somehow drift out of alignment with the other, causing the planet to wobble until it realigns its axis and spin. He considers that the earth has been rotating on its present axis for about 7,000 years, and this has caused the southern ice cap to increase its weight to 19 quadrillion tons. It is interesting to note that more scientifically temperate theories also dwell upon the hazards of the ice caps, with fears that this ice will soon melt to drown ports and coastal margins (see Chapter 10).

Surprising support for the pole shift theories has been given by the British physicist Peter Warlow in his 1982 book *The Reversing Earth*. Like other earlier writers, Warlow supports his theories with the narratives from ancient documents. If, he says, the earth tilted so as to cause China to move towards the equator, there would be a corresponding motion of the star pattern in the opposite direction, i. e. towards the north. According to Warlow this is just what a Chinese legend said did happen. The same tilt would produce the opposite effect on the other side of the globe in the Americas, pushing the northern continent further towards the Arctic. This explains, wrote Warlow, why the Delaware Indians had a Noah-like flood legend (with an individual named Nana-Bush) set in freezing cold conditions. 'The Hopi Indian legend relates that the world turned to solid ice after the sea had sloshed over the land as a result of the world spinning around crazily and rolling over', he wrote.

Geophysicists, as one might expect, view the polar flip theories with extreme scepticism. The energy required to change the earth's axis of rotation would be enormous. Shortly before Warlow's book appeared, an American physicist published some calculations deducing that the torque required to invert the earth is 200 times greater than envisaged by Warlow. This would, it seems, be too large for any cosmic body to bring about.

To the layman the concept of the crustal 'bulge' is appealing, implying that the unsymetrical elevation of huge land or ice masses could tend to unbalance the planet's equilibrium. However, the elevation of all the mountain ranges and the entire surface water put together would be insufficient to change the angle more than a few degrees. The mantle is hardly soft enough for the crust to go skidding over it. And the off-centre position of the Antarctic ice would not be enough to cause a slippage. Neo-scientists who talk about the menace of the ice caps ignore the fact that average ice cover has actually decreased this century, so in theory the last glaciation should have caused the polar flip.

The major problem with this type of catastrophe theory is one of energy. One should never underestimate the vastness of the earth's size or its irresistible momentum. Man's scientific understanding of mechanical laws is sufficiently advanced to totally discount the possibility of an earth flip. Our planet *is* like a giant gyroscope, but enough is known about gyroscopes to assert that the amount of energy needed to destabilize it and make it turn over is tremendous. Only external force could cause a change in a celestial body's rotation. Velikovsky of course argued that this was indeed what had happened. But it is unlikely the earth would survive such a knock, or even a 'fly-past'. And for the earth to reverse the spin on its axis, every fragment of matter would also have to change the orbit of its sub-atomic particles.

Is Cosmic Bombardment Possible?

Still, catastrophists continue to advance their case, and with good motives. Astronomers affirm that the greatest single threat – outside our own control – to earthly life *does* come from a missile from space. There is now little doubt that celestial bombardment was quite a frequent occurrence in the earth's distant past. Collisions with planetesimals might now be as often as one every 50 million years. And an object weighing 50,000 tons could strike the earth every 100,000 years. And, alarmingly, a 50 ton missile has a one-in-thirty years' chance of entering the atmosphere. No missile can ever tilt or destroy the earth, but some could cause quite a lot of damage to the global ecology.

Comets also can and do harbour great destructive powers. Two recent books, published in 1982 and 1983, have created a resurgence of interest in the colliding comet theory. The first was written by two distinguished Edinburgh astronomers, and was called *The Cosmic Serpent*. This title reflected the many descriptions in tribal legends of a disintegrating satellite that looked like a serpent, and was popularly associated with water. Victor Clube and Bill Napier, the authors, argued that the earth has been frequently struck by comets, meteorites, asteroids and fireballs. In this century alone at least two massive objects from outer space have hurtled to the earth's surface. Celestial impacts, they say, played an important role in evolution, sea level changes, ice ages and geomagnetic reversals. The 1908 event in Tunguska, Siberia, was a noted and much researched example, with the heat generated on entry bright enough to outshine the sun.

Even more common are luminous meteorites like the '200 ton fireball' that streaked over Europe in 1974, or the thousands of meteors spotted in 1966. Clube and Napier suggest that episodes of terrestrial bombardment by fragments of these comets instigated major catastrophes in 2500 BC and 1369 BC and 'might have caused' the Biblical deluge. Certainly, they argue, comets can be awesome spectacles, and literally put the fear of God into the primitive people of prehistory. This fear may indeed have inspired some of the more doom-laden legends of the kind that intrigues the cosmic catastrophists.

The second 'comet' book, called *The Great Extinction*, suggests that the dinosaurs were destroyed by acid rain. The cause of the catastrophe, according to the science authors James Lovelock (a distinguished atmospheric chemist) and Michael Allaby (an ecologist), was a comet 'or planetesimal' which entered the atmosphere at a shallow angle, ricocheting off to disperse thousands of tons

27

The Great Meteor of 7th October, 1868. Many scientists believe that meteors, asteroids and planetesimals of half a mile in diameter or larger can cause devastation to the global ecosystem and rapid climatic changes.

of material which was converted into aerosol particles because of the heat of entry. Volcanoes erupted and tidal waves swept the coastal areas, but it was the dust in the upper atmosphere that caused the most harm to earth by distorting the climate.

A comet, the authors suggest, could have done this if it had a nucleus about half a mile across. They say that the angle of entry into the atmosphere is crucial, as it is not the mass of the object that matters so much as the speed of its arrival – and hence the amount of energy released.

John Gribbin and Stephen Plagemann, the astrophysicists, in their recent book *Beyond the Jupiter Effect*, also make the suggestion – first advanced by the veteran astronomer Fred Hoyle – that an impact with an asteroid only 2 km across would probably be enough to cause a magnetic reversal. Impacts with comets, or even cometary fragments, could account for the major catastrophes of Biblical times. The dinosaurs, according to Gribbin and Plagemann, might similarly have been killed off by an asteroid 10 km across plunging into the ocean, vapourizing 16,000 cubic km of water. This water vapour soon condensed as snow and brought on a kind of mini ice age.

That final reference to ice brings us neatly to the subject of the next chapter.

Chapter Three

THE BIG MELT

A unique concept in those disciplines concerned with the study of the earth is that of *geologic time*. The idea is one of slowness and gradualism, and grew out of the Scottish geologist James Hutton's *The Theory of the Earth*, first advanced in 1785. The Huttonian view was regarded as an important breakthrough, usurping the hitherto dominant pre-nineteenth-century catastrophic perspective. Valleys were not dramatically and violently rent out of a benign and featureless surface, but were the product of aeons of running stream water eroding valley bottoms.

But even a cursory understanding of earth's history must temper this view. To suggest that earth processes in the past are similar to those of today does not imply that they happened at the same speed or intensity or scale. For example, weathering and erosion may have occurred faster than today. Rivers can be dated from the amount of sediment found in their deltas, and mountains can be dated according to their rate of erosion. Glaciers are dated in line with how much ground they have covered based on their present speed of retreat. But all these measurements may not present a fair and true picture of earth's history if geological forces have not always been constant.

Of greater relevance to our discussion of major historical floods is the notion that the polar ice caps may have grown or shrunk at a much faster rate than would apply today. Indeed the present-day concern with a possible rapid melting of the polar ice sheets tends to blur the distinction between the fast and galvanizing and the slow and wearing approach to geophysics. Clearly Man has suffered from as many 'uniformitarian' catastrophes as 'catastrophic' ones. Scientists today, in effect, are both uniformitarians *and* catastrophists. Continental drift theory is now largely an orthodoxy, and satisfactorily explains most geophysical phenomena such as mountain building and continental flooding. Most of these events, however, could be termed 'catastrophic' in terms of earth's 4,000 million year history.

Furthermore, a gradualist perspective would not contradict theories that the Biblical flood was the end product of an extraordinarily prolonged and devastating period of rainfall, what would be accurately known as a 'deluge'. But it is now clear that the Deluge theory is the weakest of all in explaining Noah's flood. We must look askance at the Bible's assertion that a mere forty days and nights of rain would drown the earth. Even if it rained torrentially for a whole year or more, only the very lowest reaches of mountains would be covered, and

James Hutton, 1726–1797, Scottish geologist. Hutton was one of the first scientists to advance the 'uniformitarian' explanation for earth's natural history.

the water would be pouring down the valleys to the sea at the earliest opportunity. It would be the proverbial 'drop in the ocean', and would have virtually no effect on the height of sea levels. It is interesting to note that the Egyptian account of the flood does not actually use the word 'rain'. By saying simply that water 'fell' on the land it implies that a great inundation occurred, the precise nature and causation of which is still the subject of speculation.

The theory that suggests the polar ice caps melted and caused widespread flooding is plausible since there is ample evidence that there was a great deal of solid pack ice around that no longer exists. Hence it must have disintegrated or dissolved; how fast it did this depends on which groups of scientists one believes has made the most acceptable interpretation of how earth's ice is created and dispelled. The speed of the melt would be the major determinant in the dimension of the flooding catastrophe of olden times.

Just consider the fact that at the time of the last major glaciation between 50,000 and 100,000 years ago (there have been minor ones since) there would have been as much as 18 million cubic miles of ice resting on various land surfaces: more than twice the amount existing today. In the northern hemisphere, known as the Laurentide and mainly concentrated on what is now Greenland, this ice would have covered present-day North America, Scandinavia and Europe. The water withdrawn from the oceans to feed the glaciers would, at the height of the ice age, have been some 12 per cent of the total. The unified fronts of the ice packs would have covered more than 30 per cent of the earth's surface. The glaciers would have bulldozed themselves down from Canada, flattening the forests, filling valleys and climbing mountains, until they buried even the peaks of the Adirondacks.

At the peak of this crushing era of destruction, ocean levels would have fallen by some 500 feet, and tracks of land around the margins of the seas would have appeared along what are known as the Continental Shelves. New England's shoreline would have receded by an estimated 60 miles from its present position.

But the southern hemisphere would have been more ice-bound that the north, because there is always more ice in Antarctica than in the Arctic. Right now there are some eight million cubic miles of ice resting on various polar land regions; 85 per cent is in Antarctica, 10 per cent in Greenland, and 5 per cent elsewhere. Even today the Antarctic ice cap, glistening white-blue in the sunshine, spreads over nearly six million square miles, some 3 per cent of the earth's surface, more that three miles deep in places. It is larger than Europe, and if it were centred on North Dakota it would extend from the Atlantic to the Pacific, and from Mexico to Northern Canada. To the catastrophist the growing South Pole ice cap has become a stealthy and silent force of nature, as energy created by its eccentric rotation threatens to become violently unleashed.

Hence the rapid dissolving of the last ice age would have been alarming. If the ice covering lost more than 10 inches in height per year this would account to 15 billion metric megatons of weight loss within a decade. Over thousands of years the crushing weight of glaciers must have compressed the crust of the earth to far below its pre-ice age elevations. As the land slowly rebounded during the Big Melt the pressure released beneath the surface must have been incalculable.

The earth began to fracture and contort, and much of the still depressed land was drowned as rising seas regained their old levels. It is reckoned that an average thickness of 85 feet of ice in the glaciated part of the world would be needed to make the sea level drop about three feet. So if the sea rose some 300 feet as scientists often suggest, this would imply that 1½ miles of ice melted, over a period lasting anywhere from about 200 to 3,000 years.

While this would have been imperceptible to the observer, at the very least it would have engendered extraordinary landslips and volcanic eruptions. And occasionally there would have been the sudden breaking away of huge slabs of ice to cause massive tidal waves.

The Slipping of the Ice Sheets

Peter Warlow, of *Reversing Earth* fame, is at pains to emphasize a catastrophic explanation for the Biblical flood. He points out that most geologists reject out of hand any idea that an ice melt about 10,000 years ago would cause a rapid flooding. The rate of change would have been far too slow. He tentatively offers other violent uniformitarian events such as volcanic eruptions, eclipses and earthquakes, all of which would cause an alarming darkening of the skies and/or massive seismic tidal waves, before finally dismissing them as insufficiently catastrophic.

So what does orthodox science say about the possibility of a rapid Big Melt during those times? Again we must look to the past to see how impressed were the early glaciologists – most of them confirmed catastrophists – with the rapid onset of the last ice age. The theory of the ice ages were first advanced by a Swiss scientist called Louis Agassiz. He had camped out on the glaciers of the Alps, and had made extensive geological surveys of their movements. He was convinced that rapidly moving ice sheets had had a profound impact on the shape of the earth's terrain. To test his belief he drove stakes into new glaciers, and over a period of months measured their rate of progress. He found they moved much faster than he expected, and reached the conclusion that the ice had once covered all the low ground, and had retreated to the mountains.

So rapidly did the last ice age sweep the earth, wrote Agassiz in 1840, that the woolly mammoths were soon extinguished by the dramatic fall in temperatures. earth, he surmised, had suffered many ice ages in its history, and each was terminated by renewed igneous (or volcanic) activity in the interior of the planet. So it was in the nineteenth century that we have the first scientific suggestion that the ice sheets could indeed have melted rapidly. Maybe it *was* some kind of vulcanism that caused the Great Deluge, as suggested by Agassiz, Warlow and many others.

A nineteenth-century British Professor of Geology at Oxford, Joseph Prestwich, also believed the onset and break-up of the ice ages to have had dramatic results. He was struck by phenomena that led him to believe that the south of England had been submerged to a depth of not less than 1,000 feet between the Glacial and the more recent Neolithic periods.

We have now, however, entered the somewhat contentious subject of climatic change, which is of concern as much to the astrophysicist as the climatologist.

33

We must bear in mind that throughout earth's history many remarkable things have happened to it on its relentless journey around the sun. There have been periods of rotational acceleration arising from axis wobble or tilt. The earth, too, has frequently adjusted the distribution of its mass, and it may have passed through clouds of interstellar dust. Undoubtedly these phenomena must have affected our climate. So whether scientists are inclined to support Agassiz's theories depends on whether they believe the melting of the polar caps to be geophysical or astronomical in origin: i. e. whether it relates to what happens to the earth's crust and interior, like Agassiz, or has something to do with planetary orbits. Let us take the geophysical theories first.

In the previous chapter we referred to the bizarre polar flip theories as being literally earth-shattering. But perhaps a flip would not have been so devastating if it had occurred over, say, several million years. The sheer passage of time would tend to turn it into a uniformitarian theory. And we must remember that, viewed from space, earth is not completely spherical – it is fatter at the sides, it has an 'equatorial bulge'.

In an article in *Nature* magazine in 1955, Thomas Gold, an astronomer at Cornell University, postulated that as the earth wobbled on its axis the plastic flow of the mantle would readjust this equatorial bulge. The planet may in fact have rolled over in a slow motion several times during its long history – a prolonged polar flip theory – because of the turbulence going on just below the crust. If a continent the size of South America, surmised Gold, were to be raised by about 100 feet the actual spin axis and the axis of angular momentum would cause the planet to topple over at a rate of one degree per thousand years. In the process the location of the ice pack concentrations would be shifted towards the equator. It is this – *combined* with other natural orbital wobbles and tilts – which causes the ice to melt. This, however, would be a cyclical function since the evaporation of an earlier period of glaciation would cause land masses to rise in the first place.

The idea of a plastic mantle below the crust was reiterated by two scientists at the University of Toronto in 1983. They suggested that because of the more or less rigid skin of the earth floated on a malleable but still viscous interior, the North Pole could actually change its position away from the Laurentide ice cap. So past glaciations, they imply, have shifted the axis of rotation of the earth's *skin*, rather than disturb its axial rotation. This theory is rather similar to that advanced a few years earlier by Edward Weyer. In a *Nature* article in 1978 he said that an ice age itself could be the trigger mechanism for some degree of polar slippage. Centrifugal force, under the weight of ice, he wrote, would cause the larger pole to move. This in turn would make sea levels fluctuate dramatically around the world. Weyer spoke of 5,600-year cycles, or 'rhythmic oscillations of the earth's poles'. He declared that a slippage of only one degree would raise or lower sea levels, relative to land, by as much as 373 metres.

A modified Agassiz theory was advanced recently by Richard Cameron of the American National Science Foundation. Heat, he said, radiates from the centre of the earth, melting ice caps from the bottom upwards. In addition, he argues, heat is generated from the sheer pressure of mass increasing molecular action.

As the ice cap slides outward from the centre of the polar region, friction creates further heat – a self-perpetuating effect. This is similar to Charles Hapgood's catastrophic thesis that decreed that there existed a fluid viscous layer created by temperatures in the region of 4,000°C on which the earth's crust floats. High temperatures and great pressure supposedly caused the surface to slide, partly due to the centrifugal force applied by the two centres of polar gravity.

But Agassiz's theories, after 140 years, can still hold their own. The possibility of an eruption beneath a glacier – where tremendous vulcanized heat meets with the extraordinary pressure of dense pack ice – is still a real one. The hot lava would liquify the ice for hundreds of feet over a very short time span. Devastating floods composed of icebergs as big as houses, plus volcanic debris, rocks and mud – not forgetting millions of tons of meltwater – would ensue.

Other contemporary theories suggest a sudden thaw. Recent French measurements of the oxygen isotopes of stalagmites prove that the huge ice sheets that cover Antarctica can act like a climatic time bomb. As snow continues to fall on them, the pressure at the base of the ice begins to melt much more quickly. Once its base is lubricated with water the ice can slide precipitously, with the slush acting as rollers. The dragging friction of the ground below causes further melting. Gravity exacerbates the phenomenon, causing great masses of rock and mud to move downslope.

Some argue that it is difficult to suggest any other causal agent to account for the massive infusions of fresh water into the Gulf of Mexico. According to Cesare Emiliani, a professor of geology at the University of Miami, radiocarbon dating has determined that tiny marine creatures called foraminifera died there between 11,000 and 12,000 years ago. So great was this infusion that it raised the level of all the world's oceans by 131 feet, flooding most of the coastal margins. In addition, scientists at the Australian National University produced, in 1980, firm geological evidence from the study of ancient coral reefs in New Guinea that the sea level suddenly became raised by 25 feet some 120,000 years ago. This was 'possibly' caused by a huge chunk of the Antarctic ice sheet slipping into the sea. Recently John Hollin of the University of Colorado mustered evidence from sites around the world to prove that the sea level rose by as much as 60 feet. This took place 95,000 years ago when millions of cubic miles of ice slipped off the East Antarctic ice cap, and precipitated the last major ice age (what is known as the 'ice surge' theory of the ice ages).

Another intriguing explanation for the Big Melt is the magnetic field reversal theory. We were introduced to this, too, in the last chapter. Nothing seemingly tangible takes place when the field reverses itself, but the momentary shielding effect of the earth's magnetosphere allows the planet to be exposed to much higher levels of ionizing radiation. Ionizing, or ultraviolet, radiation has sufficient energy to disrupt the structure of atoms and interfere with their electrical discharge. This could affect the balance of oxygen, hydrogen and carbon dioxide molecules in the atmosphere, and possibly increase cloud cover.

A rather better idea suggests that magnetic field reversals are accompanied by outbursts of volcanic activity. It is the volcanic dust that is said to shield much of the sun's heat, to cool off the earth with startling suddenness. There is much evidence that volcanoes across the globe had been unusually active during the past couple of million years.

The Earth in Space

The astronomical theories about ice ages draw clear distinctions between the catastrophists and the uniformitarians. The key analytical tool is the passage of, and the variability of, geologic time. Astronomically speaking, the earth is in one of its less variable periods. Its solar orbit has been reasonably close to circular for the past 30,000 years or so, but may not continue like this for the same period.

It is important to realize how massive would be the 'knock-on' effect of minuscule changes in solar radiation. If there was an alteration in the sun's strength amounting to only two percent it would be enough to cause appalling changes in the weather during the course of 50 years or more. And if there were to be a sizeable shift in the earth's orbit around the sun the effect could be disastrous for all plant and animal life.

So in the past there were probably unknown and immeasurable changes in heat and cold, and we can only guess at their biospheric impact. Stephen Schneider, a leading climatologist at the US National Centre for Atmospheric Research, believes, for example, that about 90,000 years ago a shift to near glacial conditions occurred in less than a century.

Other variables complicate the picture. The earth has a permanent roll as it rotates, and alters the tilt of its axis between being more or less upright or more tilted. The greater the tilt, the more pronounced the variation in the amount of heat received from the sun on, alternatively, the northern and southern hemisphere. To complicate matters further, the Chandler wobble occurs as the earth swivels its axis round and round over a cycle of 21,000 years. This itself might explain the extraordinary climatic sagas of Biblical times, as the northern hemisphere would have been closer to the sun in its annual summer season. Arthur Brandenberger, of Ohio State University, says the peak of the Chandler wobble, when the inclination of the axis was $26\frac{1}{2}$ degrees compared with the present $23\frac{1}{2}$ degrees, was prior to 3,000 BC. 'This date', he wrote, 'is none other than that of the Flood recorded in the Bible'.

On a longer time scale – taking up to 100,000 years – the orbit becomes stretched before reverting to its near-circular phase. At the maximum period of stretched orbit, every 50,000 years, the intensity of the sunshine reaching earth may be up to 30 per cent stronger. There can be little doubt of the climatic changes this would bring, and may have resulted in an even earlier Big Melt.

Astronomical theories are also at the root of the so-called 'heat paradox'. The surprising suggestion is made that extra solar radiation may actually bring on an ice age. This unique theory was first advanced by Sir George Simpson, a distinguished astronomer, who believed that a slight increase in the sun's heat plays a large part in getting the ice sheets to expand. All you need, he says, is much more snow in higher latitudes, and when this stays put long enough it solidifies into glaciers. Snow, as precipitation, is drawn from the oceans, and extra sunshine will cause more evaporation. It speeds up the 'heat engine' responsible for our general weather circulation patterns.

Of course, the entire temperature of the earth would be raised. But, because the earth is very nearly round, the tropics would be heated more than the poles, so there would be a greater contrast between the two regions. The ultimate effect is to speed up the flow of air and water vapour between the polar and equatorial regions. It is this that tends to create ice sheets.

Some climate observers, however, argue that this paradox doesn't exist. Robert Claiborne says that it is not true that the Equatorial Atlantic is warmer during an ice age, since isotope measurements prove that since the end of the last glaciation the surface waters have grown warmer, as common sense would predict, rather than cooler. Claiborne, author of *Climate, Man and History*, says that Simpson made the mistake of assuming that ice ages were accompanied by rain round the equator and hence an additional water vapour that could turn to snow.

Earlier a Serbian physicist, Milutin Milankovitch, advanced a modified solar energy theory. He said the sun's radiation didn't vary so much as its distribution from one place to another. It was the cool summers and springs, rather than the winters, which caused glaciation, since the winter snows took longer to evaporate – ultimately turning to ice sheets. Similarly it will be the extra hot summers, rather than the mild winters, that will bring an end to the ice sheets.

In recent years there has been much talk of the 'sunspot cycle', those dark, incandescent blobs on the sun which flare up momentarily and decrease solar radiation (they are rather like a belch, and have the effect of subtracting from the sun's output). In 1976 William C. Livingstone of the Kitt Peak National Observatory in Tucson recorded a drop of 11°F in the sun's temperature, and coincided with an increase in sunspot activity. This 11 degree fall is a change of only 0.5 per cent in actual solar energy, but if this occurred in the future and coincided with orbital changes on the part of the sun or earth, the 2 per cent ice age threshold could easily be reached. Sunspot activity reaches its peak at the end of an 11.2 year cycle, and the last sunspot maximum was in 1979. This ended what some thought was a brief but pronounced period of unstable and erratic weather.

For a long time past, but particularly in this century, some scientists have believed that the other planets in the solar system affect the solar cycle. For example, when earth and Venus are aligned on the same side of the sun the number of sunspots is said to differ. Recently this belief has become known as the 'Jupiter Effect'. This theory dictates that certain planetary alignments exert a gravitational pull on the sun, which itself in turn triggers geophysical disturbances on earth.

Another important theory suggests that the solar system as a whole passes through clouds of interstellar gas or dust, some of which is swept up by the sun. The arms of the spiral nebulae are believed to be composed of it. W.H. McCrea of Sussex University utilized the latest knowledge of spiral galaxies not only to elaborate a new theory of terrestrial glaciation, but to refine his theories of the origin of the comets and the planets. The original idea was first proposed in

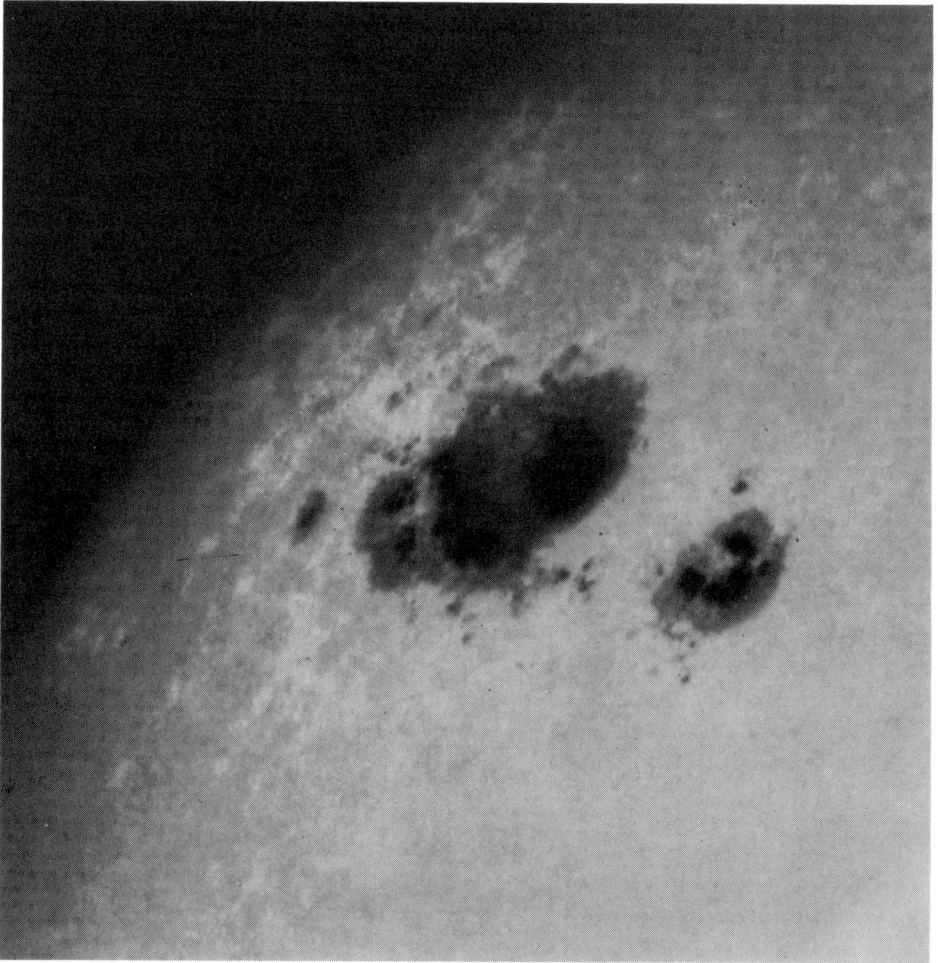

'Sunspots' close up. This photograph shows a large 'sunspot' many times as big as the earth. Very characteristic of the sunspots is their associating in pairs. They are often accompanied by exceptionally bright areas known as faculae and both phenomena are clearly visible in the photograph. It has long been suspected that sunspots cause widespread climatic disturbances.

1939 by Sir Fred Hoyle who has since elaborated upon this theme by suggesting that human life was spawned by a 'lifecloud' of dust, similar gases and dusts even being responsible for periodic epidemics of plagues. The dust yields a form of energy, causing the sun to briefly flare more fiercely. This brings about extra rain in accordance with the Heat Engine principle (see Chapter 7). In later works Hoyle referred to 'diamond dust', or ice crystals, high in the upper atmosphere which instead prevented some of the sun's heat reaching earth.

The question of climate-moulding atmospheric dust is a very contemporary one, and we will leave its further discussion to a later chapter. For our purposes here it is important to know that whatever causes the waxing and waning of the ice ages, a Big Melt would seem to occur faster than a Big Freeze. There is now general agreement among geographers that, while the water level following the last ice age rose at about a foot a century, a tremendous rise occurred several thousand years ago at a much faster rate.

Chapter Four

REMAINS OF THE LEGENDS

Neo-scientists like Immanuel Velikovksy, Hans Bellamy and Charles Hapgood – all Independent Thinkers, as Patrick Moore liked to call them – have succeeded in irritating the probing, measuring world of the earth scientists, or those who bothered to take any notice. These writers postulated pulverised lands ground into the earth and covered by tidal waves in the space of a wild season of speculative horror. And all while the established laws of astrophysics were momentarily held in abeyance.

Yet in many ways they were advancing theories no more extraordinary than Alfred Wegener's Continental Drift theory of 1915, but which was largely rejected until confirmed by Sir Edward Bullard's computer in 1964. It was then that heresy was soon turned into scientific orthodoxy. The Velikovskies of this world were writing in the dark ages of geology, prior to the revolution in the earth sciences in the 1960s, since when a much better understanding of the earth has come about.

One might easily suggest that it was the nomenclature of the neo-scientists that was largely at fault. Their error was to refer to Atlantis while others were referring to the Mid-Atlantic Ridge. Ignatious Donnelly, in addition, was referring to that other long-lost continent Lemuria when geologists were writing about Pangea. And whereas the German biologist Ernst Haeckel suggested Lemuria was populated by lemurs (thus giving Lemuria its name), during the Cenozoic age, Donnelly said it was populated also by humans.

And it hardly seems fair that Velikovsky should be criticised for interpreting ancient writings as some form of meteorological notesheet. Few anthropologists deny that climatic vagaries have been historically important to the human race. We know that ice ages existed, and we know that pre-glacial Man existed (although we are far less certain about his attainments). Indeed many writers point with pride to early Man's tough origins in the dim and freezing mists of the ice ages, with its implication of stoical endurance, of honing trials of strength against fearsome wild creatures.

The early nomads of these days would have found it more congenial to live on the newly exposed coastal regions, as far away from the chilly hilly regions as possible. The lowered sea shelves would hence have become communal meeting places, where humans from all over the hitherto inhabited world would exchange primitive knowledge and cultural values.

This might provide a clue to the idea of cultural diffusionism, and explain the similarity of the universal Flood legends. The sea, in effect, as well as being the cradle of life, was also the cradle of civilization. More importantly, life on the sea floor meant that the waters would henceforth become the perennial and mortal enemy of mankind, with the collective experience being one of retreat to higher ground before the onrushing tide.

But before the Big Melt, and after the halt of the growth of the ice sheets, permanent settlement would have been possible for a new race of people. Conditions on the sea floors would have made life easier than it had been for their ape-like predecessors eking out their grim existence on the frozen tundra away from the equatorial regions. The soil, for example, sheltered from the worst ravages of the ice, would have been excellent.

Frank Walworth, in his controversial book *Subdue the Earth*, claims there were at least three great centres of civilization on the seas' littoral shelves. One would have been in the Atlantic Basin on either side of what is now the Mid-Atlantic Ridge, which would then have been above the waves. Another would have been in the eastern Pacific, possibly centred in the vicinity of Easter Island. There would have been another in the Indian Ocean Basin off the coast of Africa. Walworth says there may have been other settlements, all scattered among plateaux hidden among fertile valleys and shallow seas fed by the glacial rivers.

Climatic conditions would have remained static for perhaps a few thousand years. Then, as the earth began to warm again, the ice sheets would begin their fateful shrinking. With the onset of the Big Melt there would have been distinct increase in wetness among many coastal areas during the lifetime of one long-lived adult. The ice age would have become a 'Water Age'. The rising seas would have covered large portions of the continents, reducing them to low islands separated by shallow gulfs and bays.

Throughout the physical world layered strata show that shorelines disappeared beneath the water, or instead were forced upwards by enormous tectonic forces. Caves and grottoes below the sea show vital evidence of stalactites and stalagmites which can only be created by the evaporation of mineral-bearing waters in the open air.

Several hundred miles to the north of the Azores and to Gough Island in the south lies the mountainous Mid-Atlantic Ridge that roughly parallels the African coastline, like a subaquatic shadow. The loftiest peaks of this range still exist today in the form of the Azores themselves, Ascension Island and Tristan da Cunha. Certainly the Ridge looks like a backbone of mountains that once joined the two Supercontinents.

To complicate matters it is located on the world's most active earthquake belt, with constant volcanic disturbances upheaving the topography of the ocean bottom. Within the past 1,500 years, it is reckoned, islands in the region have appeared and become submerged, and have grown and shrunk in size. And the Azores today are said to have unexplained animal and insect life, implying they

41

were indeed once part of a continental land-bridge. One of the Azores islands, Pico, is a large mountain some 24,000 feet high, of which 16,400 feet are under water.

In 1949 a team from Columbia University explored the Ridge and were said to have found sand in its upper shelves. And as sand is the product of open-air erosion, some obvious conclusions were arrived at. Some years later P.W. Kolbe, of the Riks Museum in Stockholm, discovered tiny shells of miniscule freshwater creatures that could only have been deposited in the sediment of what was part of a freshwater lake. Another expedition, this time sponsored by the Russian Academy of Science, dredged up rocks in the region of the Ridge which seemed to have been formed under atmospheric pressure, and to have been submerged around 15,000 BC. The Bermuda Islands, some 2,000 miles to the west, are reputed to be the peaks of immense underwater mountains which were above sea level during the last ice age. But the most important recent findings were made by the University of Miami marine scientists in 1975. Fossils and limestones taken right off the surface of the Atlantic sea floor actually show traces of rainwater. And they also believe that a sudden flood of icy waters caused deep-sea marine life to drastically alter their characteristics because of some event that occurred 11,500 years ago.

With this knowledge in mind we can begin to perceive how the Atlantis legends might have originated, assuming that they have any basis at all in reality. For at the height of the Big Melt most oceanic islands 'sank', and land-bridges and even mountains were submerged by between 100 and 500 feet.

But this doesn't make the search for Atlantis any easier, since most of these islands would have been large enough to support a primitive civilization. The Atlantic Ocean, in fact, possesses numerous legends about great civilizations among mysterious and scattered islands called the Fortunate Islands, Islands of the Blest, or Hesperides. The problem is that any sizeable sunken land mass would not have been the product of the *last* ice age, but earlier ones, dating back millions of years. It may even have arisen and become destroyed in the geotectonic upheaval that took place shortly after the planet came into being.

And yet if one assumes the catastrophe to have taken place at a time closer to our own, it is possible that the Atlantis legends relate to the Mediterranean island of Crete, once said to be ruled by King Minos. We know for a fact that the Cretan civilization was highly developed, even if some doubt is cast upon the existence of Minos himself. About 80 miles away from the island lies Thera, which is a volcanic island. Archaeological evidence reveals that the volcano erupted with astonishing violence around 1500 BC, and the effects upon Crete were catastrophic. The coastal regions of the island were swamped, and there was a very high death toll. The so-called Minoan civilization never recovered, and eventually disappeared from history.

In 1883, in Krakatoa, East Indies, a volcano behaved in a similar manner, destroying the island on which it was located and flooding with huge tidal waves the coasts of densely populated neighbouring islands. The shock waves from this momentous event even affected the English Channel, on the other side of the globe. Over 30,000 people lost their lives, and the dust arising from the eruption

42

lingered in the upper atmosphere for three years. The western seaboard of North America, in whose waters are supposed to reside the remains of Mu and Lemuria, also has a crust which is a notorious fault-zone. For millions of years earthquakes have ravaged the area.

Floods and the Decline of Nations

Whatever one may believe about the Atlantis and Great Deluge legends, great floods in the past are not mythical. The evidence shows they were real and disastrous, inspiring much apprehension and insecurity. The early civilized Greeks, the pioneers of philosophical discourse, were yet fearful of the 'fire and water' that periodically destroyed their culture, and which historians coming across written references to such fears had thought were remnants of earlier 'mythical guises'. Climatologists Reid Bryson and T.J. Murray cite the myths of Homer in their book *Climates of Hunger* which relate the stories of Troy and Mycenea, the remains of which have actually been traced and identified. Historic cities often suffered from modern-day natural catastrophes, as well as others peculiar to their own time.

In more senses than one water for mankind has always been one of the most serious concerns. And climate, as the major determinant of this life-sustaining liquid, is likely to have played a decisive role in human history. So much so that early, primitive peoples at the mercy of the elements felt compelled to appease their gods. Pre-literate tribes in New Guinea and other places still cultivate the land, grow crops in seasons, and use flint and obsidian for making primitive tools. Yet, schizoid-like, this is felt to be insufficient, for they still perform grotesque dance rituals to appease the awesome destructive power of the wind, rain and the sun. Even the civilized Egyptians had their god of inundation, Hapi, who was loved as well as feared for his capriciousness for preventing the drought, but never knowing when enough was enough. And in spite of Hapi, the Egyptians went on irrigating their arable lands, to apply primitive technology as an aid to survival.

The pathological, unreasoned way in which weather patterns manifest themselves gives rise to strange human responses. Some archaeologists and historians are now being obliged to revise old theories about the decline and fall of certain empires to place more emphasis upon climatic factors. In 1915 the American philosopher Ellsworthy Huntingdon wrote that civilizations tended to rise and fall in accordance with a 600 year climatic cycle. He wrote of 'nomadic eruptions', such as the sudden rise to prominence of the Vikings 1,000 years ago.

There is now a tendency to explain the vissicitudes of the African civilizations in terms of the prevalence of drought, and that of the Chinese civilization on the frequency of floods. Even the barbarian invasions in Central Europe and Asia may not have been entirely due to abstract concepts of militarism and conquest. There are usually pressing environmental reasons why peoples feel compelled to push into adjacent territories, usurping any indigenous populations in their way. This may have applied to the Indus civilization, and to the Hittites as well as the

Mali civilization of Africa. The American writer, Lowell Ponte, says that the greatest military achievement of Rome was the conquest of Egypt in the first century BC, then the food-basket of the Mediterranean. Thus the Caesars 'could take men from farming and put them into armies, and thus they could conquer the world'.

And yet the limits to territorial expansion was also determined by climate, to the margins of severe frost in the north and the deserts in the south. One thousand years after the collapse of the Roman Empire the cooling climate tempted several European nations to try to build empires of their own. This new imperialism, according to Ponte, should not automatically be ascribed to mercantilism or greed so much as to the severe cold climate that had returned to Europe in medieval times and brought about many famines.

The distinguished British interwar meteorologist, C.E.P. Brookes, said that floods in the Bronze Age (*c* 5,000–1,000 BC) may have some connection with the sudden appearance of the Phrygian peoples traversing the Hungarian Plain soon after 1,300 BC, and heading towards a then drier Asia. About a thousand years later the trend was reversed, with the thrust into the western world by the nomadic Mongolians. H.G. Wells, in his concise *Short History of the World*, postulates that climatic change was the trigger. He leaves open for speculation the idea that reduced rainfall abolished the swamps and forests, or greater rainfall instead extended grazing possibilities to desert steppes to which the Mongolians had to migrate. Other examples of climatic compulsion are provided by one of the world's top climatologists, Hubert Lamb, in his recent *Climate, History and the Modern World*. He refers to the flooding of prehistoric lake villages in Central Europe around 800 BC, as well as variations in moisture levels and forest growth in the valleys of Mexico and the Yucatan, and Cambodia, hinting at serious flooding.

The Great Migrations

Man's fate was to be born into earth's ecosystem at a time of rapid climatic change. Fifty million years ago the earth's topographical features would have been similar to today's. But the vegetation would have been different. Then palm trees grew where Paris is now, and cinnamon and mangos flourished in parts of Germany. Temperatures were then well up on those prevailing at the present. The poles were milder, and the temperate regions of Europe were subtropical, and in places tropical.

From that high peak things slowly deteriorated. The tropical vegetation in the northern hemisphere began yielding to subtropical species like cypress and magnolia. A few million years ago these in turn had given way to vegetation much like that of today's. Certainly by about 10,000 years ago, at the start of the Biblical era, the biosphere – the earth, its atmosphere and its ecology – was almost identical to our own. But there was much more forestland; oak forests in Macedonia and the Mediterranean region flourished. The various land strips connecting the larger land masses had probably already been eaten through. The great barrier across the Straits of Gibraltar had sunk, as had that between the Flemish Bight and south-eastern England.

The archaeological evidence, from this point on, from the Atlantic littoral shelves and elsewhere, suggests, as we have seen, that Caribbean civilizations evolved at a time when ocean levels were at their lowest. After that, land areas were submerged when the Bahama Shelf was inundated after the melting of the northern glaciers. The flooding is likely to have been very gradual. Many gargantuan submerged walls appear to have been dikes built in a vain attempt to protect certain areas from the rising sea. And as the sea-bed tribes around the world retreated from the rising waters, they would disperse in different directions.

In the Atlantic Basin some headed eastwards towards Europe, some westwards to the Americas, and others to Africa. Some would have climbed up the Mid-Atlantic Ridge to attempt to survive in what is now the Azores. They all became the first primitive nomadic peoples. Scientists from the Arid Zone Research Institute in the Negev, Israel, have pointed out that the world's first city, Jericho, coincided with the beginnings of the rapid rise in sea level over the coastal plains, and that the inhabitants took advantage of the rich salt deposits of the nearby Dead Sea.

Indeed, a watery environment was considered highly desirable. A glance at a map of ancient mounds and village sites shows the rivers, lakes and springs around which historic settlements have been established. Foreign invaders immediately settled in the most fertile deltas, as the invading Romans did in London in 55 BC. And in the new world a variety of Indian civilizations settled on the banks of the Colorado, the Amazon and the Mississippi.

The early great civilizations, then, appeared in what is known as the Middle East and Asia. The descendents of the earlier postglacial nomads eventually populated three major regions where civilization started and flourished. These were the Indus Valley (present-day Pakistan and India), Mesopotamia (now the region of Iraq), and later China. All three were situated around great and volatile river systems: the Indians had their Indus, the Mesopotamians their Tigris and Euphrates, and the Chinese their Yangtze. Other civilizations in Europe congregated around the Danube, the Rhine, the Po, Seine, Volga and the Thames, all of which shaped the face of Europe when dense urban areas were eventually built around them. In other regions the world's early populations lived on coastal plains, far more so than they do today, because they were fishing and maritime communities, and they needed the evaporating salt water to preserve food caught on the land.

Babylonia, before it became Mesopotamia, was the land of the Sumerians who gave the empire its civilization and culture. The Sumerians are credited with opening up the Tigris-Euphrates alluvium which was, some 8,000 years ago, a virtual jungle-swamp. The Aegean races lived near large inland seas in central Europe and on coastal plains and small islets. Memphis and Thebes were important cities located on the Nile Delta. Nineveh was on the Tigris, Sidon and Jerusalem were on the eastern edge of the Mediterranean, while Carthage was at the western end. Persopolis was near the Persian Gulf. All around this central Asian area stretched moist, fertile and habitable lands, much of which today is parched scrubland. But in those days there was a handicap of excess surface

45

moisture rather than the lack of it. Even European Russia was more of a giant swamp in those times, and those swamps possibly separated the Nordic and Hunnish peoples that H.G. Wells wrote about.

Immense difficulties were often encountered with managing the surface water, and the springs, rivers and lakes. Richard Barnet, in his *The Lean Years*, has pointed to the intricate network of canals and sewers that the Sumerians developed. Their civilization literally disintegrated when these life-support systems fell into disrepair. The Chinese, being very early victims of floods on a massive scale, were fortunate in having the benign ruler Yu. Yu mastered the art of flood control as early as 2300 BC, and inspired a high culture sustained by irrigated agriculture. He established a dynasty that lasted 800 years.

But why were these regions so much wetter than they are now? Firstly, archaeologists believe that within the last 4,000 to 6,000 years the Caspian Sea in Asia Minor received more than its fair share of ice melt. This is because it is a closed sea with a smaller ratio of water in comparison with the surrounding land. Not being part of the vast ocean storage, the shrinking glaciers of the Caucasus and the Ust Art Plateaux regions would have found a ready catchment area.

There would also have been widespread flooding and raised soil-water levels in what is now Armenia, Iran and Turkey. It is even possible that at one stage the Caspian Sea joined up with the Black Sea. The Greek islands in the Aegean Sea would have been all but submerged. The Tigris and Euphrates would have been in full spate for generations. Indeed, Robert Claiborne, in his *Climate, Man and History*, says that the Euphrates carried more silt even than the Nile. Over thousands of years its bed and banks have been built up well above the surrounding plain. Water for irrigation was often obtained by simply gouging out chunks of the levee, which all too soon crumbled away in flood time. The flood-waters would spread for miles. This happened so many times it might well have seemed to have dominated the world, thus giving rise to the Deluge legends.

Secondly, there are the important geotectonic theories. German and Swiss scientists as early as 1923 found that at around 6,000 BC strong landslip movements in the Bavarian Alps caused glaciers to melt rapidly and lakes to overflow. Similarly George F. Dales, the archaeologist, believes that the Indus floods during the Harappan period resulted from earth movements which periodically hurled dams of rock or mud across the lower Indus, thus inundating the valleys below. There is a lot of evidence to show that the Harappan cities, around 5,600 BC were abandoned possibly due to invasion or destruction. Thomas Wigley, editor-contributor to *Climate and History*, also agrees that drastic tectonic movements caused floods in this region. Perhaps, he suggested, the Indus River underwent a major, abrupt change of course.

Searching for the Proof

The most important technique in searching for past floods is to examine sediment and other geological strata, such as silt, sand and gravel. There is the evidence of dried river channels, which give important clues as to the balance

Charles Leonard Woolley, 1880–1960, archaeologist and writer, led an expedition to excavate the tombs of Ur in Iraq.

between rainfall surpluses and evaporation. The surrounding soil sometimes reveals similar sedimentary deposits, indicating that the river frequently over-flowed.

In 1923, the year when Howard Carter first peered into Tutankhamun's tomb and saw 'strange animals, statues and gold', archaeologist Leonard Woolley was to achieve fame through his joint Anglo-US digs at Ur, that fabled and ruined city of the Chaldeans. Six years later Woolley and his team completed their excavations. They had cut a 62 foot section through the earth which showed not just evidence of human occupation but also a 10 foot stratum of clay or sand. In this they discovered remnants of tiny marine creatures, but none of human life.

The clay and sand stratum, they surmised, could only have been created by water, though its thickness remained a mystery. A normal accumulation of a ten foot thick deposit would have taken a long time. But the strata was sandwiched between layers where remains of similar civilizations (at different levels of accomplishment) were found. So it appeared that a large deposit of silt was laid in an extremely short time.

Many earth scientists, however, have been out of sympathy with Woolley's claims that his findings confirmed the Genesis story, which the press of the day sensationalised. Even so, Woolley's claims in his 1963 book *Excavations at Ur* are mundane, merely bearing out some of the more prosaic descriptive details in Genesis. 'The Sumerian version of the Flood', he wrote, 'describes antideluvian man living in huts made of reeds, which at al'Ubaid and Ur were found to be the case: Noah built his ark of light wood waterproofed with bitumen . . .'. He goes on to deny the universal Flood thesis, saying it effected only the Tigris and Euphrates area; 'for the people who lived there that was all the world'.

Many other experts now believe the 'legendary diffusion' theories of the Flood. They point out that Ur preceded the Babylonian and Sumerian civilizations. The accepted view was one of rapid climatic deterioration. Similar deposits at Kish and elsewhere in Mesopotamia have yielded a range of dates between 4,000 and 2,000 BC, which seems to have been wetter than at any time since. But the Kish findings succeeded only in turning the Great Flood legends into a Deluge legend.

In fact there is much evidence for a combined rain and sea inundation in the Tigris/Euphrates Basin when possibly seismic phenomena caused tidal waves in the Persian Gulf. Soil researches in the 1970s in the plains of Iraq and the Gulf, and the surprising quantities of silty clay cores dragged up from the floor of the Gulf of Mexico, have hinted strongly at disturbingly serious floods in early historic times. Take the evidence provided from boreholes made by the Iraq Petrol Company. A great many freshwater fossils were uncovered from the thick alluvial clays and sands around the village site of Dar-i-Khazineh, north-east of Basrah. The village itself, occupied from 6,000 to 7,000 years ago, rested on a body of alluvial deposits some ten feet above the level of a previous settlement. In the January 1982 issue of *Nature* a team of French experts also suggested that severe flooding in the Nile Valley occurred between 8,000 and 9,000 years ago. This is said to account for the distinctive layer of saprosel mud deposited in cores taken from the bottom of the eastern Mediterranean Sea.

Finally there is the evidence of climatic change. We know that by 5,000 BC the temperature was warmer than it had been for 100,000 years before. Europe was then eight or nine degrees F warmer than it has been since, reaching what is known as the Climatic Optimum. West and Central Asia suffered from abnormally severe storms as the rainfall belts shifted dramatically. Excessive rainfall in the rest of Europe swamped many forests and turned them into peat bogs.

There were marked rainfall maxima at about 4500 BC, and then again at about 400 or 500 year intervals, with lesser periods of rain at about 200 year intervals, until the time of Alexander the Great (356–323 BC). The worst flooding era in Europe, Asia and America was likely to have been from 2400 BC to 1500 BC, and then again at the time of Alexander. A great number of towns and communal settlements were submerged under enormous depths of flood silt. Some villages were often completely rebuilt on the ruins of their predecessors. In Scandinavia it was so wet around 350 BC that boats were used to travel from one village to another. The British Isles were also so marshy and swampy that the earlier settlers had to use wooden trackways to cross the fens and marshlands in Essex and other flat lowland. Later the trackways had to be abandoned in favour of boats. Villages were actually built high up in the middle of lakes to minimise the effects of further storm floods.

In more senses than one the arrival of Christ on earth marked the end of an era. Later generations would discover both the awesome legends that purported to describe the Great Floods, as well as the physical remains of those real events. But those same learned generations would soon suffer the fearsome floods of their own time. Throughout history, and right up to the present, mankind continues to fall victim to sinking lands, catastrophic tidal waves and great deluges. The Flood legends were not meant merely to be histories, but to be warnings.

Chapter Five

FLOODWAVE!

One cannot help but remain surprised at how earth, alone in the solar system, became a wet planet. Our two nearest neighbours, Venus and Mars, are in a condition that earth must have been like shortly after creation – small, void and rocky – but they are still lifeless today.

It was water that first created the vegetative life which in turn generated the oxygen necessary to break down the carbon dioxide. It was this gas (rather than methane and ammonia as is commonly thought) that was probably the dominant gas on earth before the primeval rains came. And it was this new life-giving atmosphere that enabled animal life to appear for the first time.

Nevertheless earth's habitable biosphere is entirely accidental. Our planet is roughly in the middle of the band of orbits around the sun where a rocky planet can maintain a life-forming environment. In between the absolute zero of interstellar space and the millions of degrees of the flaming stars, earth has an in-between temperature that permits flowing water, and hence a stable atmosphere. Any moisture vapour on Venus would be vaporized into space as it is too close to the sun. True, Venus is shrouded in clouds that create a tremendous greenhouse effect, but these clouds are now believed to consist of a dry and lethal mixture of sulphur and carbon dioxide.

On the other hand, whatever moisture there is on Mars would soon turn into solid ice, as the planet is 47 million miles further away from the sun than earth. Mercury is not only a virtual celestial cinder but, unlike Venus, is so small it lacks sufficient gravity to hold an atmosphere or gases such as water vapour.

The giant planets that lie beyond Mars all have surface temperatures ranging downwards of 130°C below zero. Ice probably makes up the outer layers of Jupiter and Saturn, and possibly those of Uranus and Neptune as well. Saturn's rings are believed to consist of a swarm of ice particles, probably remnants of a shattered moon.

But how, exactly, was earth's water created? In the distant mists of time when the earth was being hewn out of plasma and cosmic debris and squeezed together by gravity, water vapour and gas was forced out of the amorphous and gelling rocky material. As the condensation of the heat, too, caused moisture to evaporate it was first simply vaporized high into the airless void above. Gradually this gaseous moisture began to draw some of the heat from the surface, and a few drops of rain actually fell to earth.

Then the start of the most momentous epoch of earth's development took

The Wet Planet. Earth's hydrosphere becomes obvious from space. The dense clouds, the massive oceans, the ice packs and turbulent weather patterns are a constant menace to mankind.

place. The rains increased inexorably, and the land beneath the smouldering surface cooled further, and soon torrents of primitive water, more like solutions of gases, were falling. Hot streams of muddy liquid poured over the crystallizing rocks and lava to carry detritus and deposit sediment. Violent winds, as hot as furnaces, were accompanied by the downpour, to cut deep gorges and canyons, and to dump debris into the early turbulent seas under a lowering sky dense with clouds.

Still the earth was quite lifeless. There were crags and hills of barren rock, but no soil or plants to break up the rock particles into mould, nor was there lichen.

51

Finally the rains ceased, and the storm-rent skies began to clear. The seas swelled and the rivers flowed, and a liquid outer layer of the earth had been formed. The now earthbound water continued with the moulding of the crust that the Great Rains had done earlier. The water defined the first continental blocks, slowly clearing channels and whittling away at mountains and sweeping their remains into marine basins.

The life-forming water also managed to control the earth's surface temperature. Our planet, with its newly formed hydrosphere settled down to an average temperature of 15°C, down from an average of 25°C at the beginning.

When the oceans were at last full they comprised about 97 per cent of all the water on earth, with freshwater lakes and rivers making up the other 3 per cent, The total surface area covered by all the world's oceans is nearly 140 million square miles, or some 71 per cent of the earth's surface. The quantity of oceanic water represents an enormous 325 million cubic miles.

Certainly the hydrosphere is thin in comparison with the lithosphere, the solid ball of the earth itself. And yet if it were accumulated into one place it would form a sphere about 864 miles in diameter – probably larger than all the asteroids or planetesimals in the solar system put together. If the oceans were divided up each man, woman and child would get 110,000 *million* gallons.

So there is plenty of water, and it is spread out across the globe. If the lithosphere was completely spherical the entire planet would be uniformly covered in water to a depth of almost 8,000 feet. But our planet has marked irregularities. Not only is the Pacific almost as large as the other oceans combined, at almost 14,000 feet deep it is the deepest. It also means that the southern hemisphere of the globe has 81 per cent of its surface covered by water. Whereas the Pacific represents nearly half of the world's ocean, the next biggest sea area, the Atlantic, takes up 30 per cent of it. The Indian Ocean is third in size, comprising – with its 30 million square miles – some 21.5 per cent of the world ocean.

Our Sinking World

Some time during the Jurassic period, about 160 million years ago, earth suffered its greatest flood of all. It was a disaster that endured for a long time – in fact about 45 million years – but it was of momentous significance for the future of every living thing on the planet. For during this period the seas were deeper and higher than they have ever been since. They indundated great portions of the globe: the western parts of North America, Europe, Siberia and most of Africa. Then, the earth was sinking. And it has been sinking – and rising – incredibly slowly ever since.

Now, thanks to the rapid advances in geophysics in recent years, we know that this slow-motion turbulence accounts for the baffling distribution of the fossils; of the hippo and bison bones found in the soil of Britain, and of the mastodon teeth dragged up from the east coast of the United States. A sinking world would also explain why peat, formed from decomposed vegetation in marshy ground, has been found on land presently high and dry. And why, for example, fossiliferous limestone, formed in warm shallow seas, has been discovered several thousand feet above sea level, even close to the summit of Everest. It would explain why animals and humans could cross the thin Bering Sea to Alaska right up to most recent times, and why parts of the bed of the Atlantic are made of tachylyte, a vitreous basaltic lava that only cools above water.

Ironically it was the discovery of the sideways drifting of the continents that also helped explain the downward shifts of the earth's crust. From the sixties onwards a patchwork knowledge about earth suddenly assumed a wider, coherent frame of reference. The theories of Georges Cuvier, the great French paleontologist, now have more meaning. He was one of the first to draw attention to the symbiotic relationship between the land and the sea. From very early times it was known that the sea changed its level from time to time because of the disconcerting discovery of ruined buildings on the sea floor. What else could explain ancient harbours with moorings for boats found high up on dry ground some distance behind?

To get some idea of why the solid and liquid components of the geosphere – earth and its atmosphere – seem to be perpetually at war, we must look briefly at earth's natural history. The conventional wisdom runs something like this: about 4,200 million years ago, after the surface of the earth had cooled, there was one huge land mass called Pangaea ('all earth'). Then the first seas originated in depressions in the land, and then grew deeper as erosion took its toll. Soon there emerged a large ocean called Panthalasia ('all sea').

It was the enormous stresses within the lithosphere which is supposed to have wrenched Pangaea into two segments separated by Tethis, a vast primeval Mediterranean ocean. The northern mass was called Lurasia, and comprised the present North America, Europe and North and Central Asia. Gondwana was the other half, and consisted of South America, Africa, India, Antarctica and Australia. But it was the formation of the hydrosphere that was of vital importance to the destiny of earth. It stimulated the growth of plant, and later animal, life, and had a more equalizing effect on global temperatures as a moist atmosphere carried cooling, life-giving precipitation across the globe.

Ice: The Earth Shaper

At this stage we must introduce the role of ice into the history of the earth's formation, as it is already clear how important it is to the subject matter of this book.

According to very tentative evidence, there were at least five ice ages in Pre-Cambrian times, i.e. during the period ranging from 4,200 million to 570 million years ago. The latest geophysical reasoning says that there have been many more than this in Post-Cambrian times (i.e. from 570 million years ago to the present, encompassing three major geological eras). Some scientists, using differing definitions of the 'ice ages', say that there have been even more 'ice epochs', separated by 'interglacials', in which occur innumerable ice ages and mini ice ages of varying severity. The Czech geologist Josef Sadil, for instance, believes there were no less than thirteen glacial periods of greater or lesser severity in Europe during the Quaternary period, which is during the last three million years to the present.

Shortly after the creation of the giant Tethis Sea, part of the northern supercontinent overlapped one pole, allowing ice to form and create probably the first great ice age. Then with the passage of time, the tremendous subterranean forces and the great mounds of pack ice (unrelieved by any equatorial flowing seas), eventually cracked Gondwana into fault zones. And when Pangaea also began to break up, the warming ocean currents facilitated the first series of ice meltings. South America finally began to drift apart from Africa, leaving a narrow shelf with a string of offshore volcanic islands.

The most important result of the ice ages was the additional pressure they caused to the crust of the earth. The sideways slippage on a viscuous magna was compounded by the weight of the ice as it succeeded in folding or lifting the crust into great mountains.

The beginning of a curious cyclical geospheric process forced its logic on the earth. First the ice ages were caused by changes in the sun's radiation, then waxing and waning of the ice caused mountains to form, and these in turn played their part in getting further ice ages started. Eventually the mountains were gnawed at by the elements, and the sediment created by the erosion was washed into the sea. This raised the ocean bottom and caused the waters in turn to flood back onto the land, and so on.

We can also begin to understand how the ocean and land masses change places with each other. In the twenty years since the acceptance of Continental Drift, many scientists began to see a crucial link between drift and sea-floor spreading. It was Robert Dietz, a marine geologist at the Atlantic Oceanographic and Meteorological Laboratories in Miami, who developed this hypothesis. It was suggested that new ocean floor material wells out continuously from cracks running along the edges of mid-ocean ridges, and effectively pushes the ocean bed further and further apart. Soon a single unified concept, plate tectonics, was advanced. The earth's outer shell was thought to be split up into large and small rigid plates which interact with each other by growing, destroying and slipping underneath each other at the margins. This raised and lowered the ocean in the process.

The Final Shape of the World

Then, in the moist and rainy Carboniferous period (330 million years ago), when ferns and club-mosses provided our present-day coal deposits, the Tethis Sea spread in two branches, one westwards to the coast of Canada and the other to West Africa. In the east the new sea reached India on its way through Central Asia, splitting into two smaller seas to head towards the Bering Straits in the north, and reaching southwards to Indonesia and Australia.

After the Permian Ice Epoch ended about 240 million years ago the poles were surrounded by ice-free oceans, and this caused the earth to warm. This was suggested in 1956 by Maurice Ewing and William Dunn of Columbia University, who said that the warm ocean currents could travel up from the equatorial regions. And with further plate shifting the Arctic Ocean was cut off and the tropical ocean currents were severed. But the warming continued until the start of the Mesozoic era, to culminate in the prolonged rise in sea levels in the Triassic and Jurassic periods.

Finally, as the Mesozoic era drew to a close, something else happened that was to have a great impact on the flooding scene: definite zones of climate appeared. From now on the perpetual interaction between the tropic air flows and those of the cooler, moister regions would mean that the world would henceforth suffer virulent storms and torrential rains.

It was from then on that the earth seemed to become a much wetter place. In the Cretaceous period, about 80 million years ago, Tethis linked up with the Antarctic Ocean via the South Atlantic and Indian Oceans. There were catastrophic sea floods which inundated huge segments of central and eastern Europe and the newly formed Volga River, as well as the western tip of Europe, paving the way for the break-away of the British Isles.

In the Paleocene epoch, about 45 million years ago, Europe began to sink once more, and the sea encroached for hundreds of miles over a 10 million year period.

The seas also expanded in Africa, Asia and America. Western Europe was reduced to the status of one of many large islands, and virtually the entire Middle East was under water. While India was an island lying off the Asian coast, a great tranche of seawater split off Northern Europe from Siberia.

From much of the sedimentary evidence of these frequent incursions we can speculate that the ice sheets were again melting. There is considerable proof of a rise in temperature. In fact Europe and America had their warmest times since the very formative days of the earth, and there was much lush sub-tropical vegetation. Britain at last began to separate from Europe, with the English Channel finally appearing about 7500 to 5500 BC, i.e. very recently, since it was possible for primitive Nordic tribes to wander from Europe to Britain on foot.

In the Quaternary period, about three million years ago, Tethis had metamorphosed into smaller basins such as the Mediterranean. As the crust continued to slide on its magma, and the hitherto yawning ocean from prehistoric times was more and more confined by the land, local regions continued to be disastrously flooded. The middle west of the United States was frequently submerged, as witnessed by the famous drowned riverine system. A remarkable 2,500 miles long, it had many tributaries, one of which was probably the Hudson River. This subaquatic river is said to continue on the sea bed through underwater cliffs to join up with the Ridge after passing the Sargasso Basin, east of Florida. Similar submerged river systems extend out into the present Atlantic from the shores of Spain, Portugal and France.

Is the world still sinking? Most probably, yes. Some scientists believe the oceans are growing steadily. The geophysical processes that shaped the earth are still at work, and volcanoes continue to pour water vapour into the air to continue a process that started with the outgassing of primeval hot rocks. As earth continues to contract under its own weight this vaporization will continue remorselessly. And the weight of the water will go on forcing the ocean beds downwards so that the oceans get deeper and deeper.

No Enemy but the Sea

The importance of the seas has long been etched into the consciousness and imagination of the human race. The Sumerians and Babylonians thought of the universe as an immense ocean that continued from the horizon, through various glittering shades of translucence, way into the sky. The earth, like a huge circular and flat island, had somehow emerged from the middle of this celestial sea. Homer tells of Oceanus, who was the creator of the universe and the gods, the human race, and much else. Thus were the beginnings of an appreciation of the dominance – both in real terms and in practical day-to-day terms – of the oceans.

As the early civilizations settled around the world's great river plateaux and near large inland seas they would soon have developed a tolerance to the threat from the watery sector of the earth. Those living in coastal regions had the most to fear. The waves and currents relentlessly attacked the land, eroding here and depositing there, with the occasional tidal wave demolishing whole fishing communities that had become established too near the coast.

For primitive communities by the sea the expense of coastal defence at periods of rapid aquatic invasions – such as at the time of the Big Melt – would have been astronomical and probably futile in the long run. Barriers even 150 feet high would sooner or later have collapsed, and presented an unacceptable risk to those living by the increasingly turbulent sea.

Bit by bit the evidence of this struggle against the waves comes to light, sometimes inadvertently but mostly, as we saw in the last chapter, through professional investigation. Geological studies from the island of Walcheren in Holland have shown that Roman settlements, established about the end of the first century AD on a clay ridge in an area of peat, were entirely engulfed by the penetrating sea towards the end of the third century AD.

The British Isles, in particular, have long been at the mercy of the rampaging seas. Human artifacts, and the bones of land animals, have been dredged from the bottom of the North Sea. Norfolk fishermen have discovered a spearhead carved from a deer's antlers, and dating from the early Neolithic Age. In 1878 Roman artifacts were uncovered in the Essex tidal mud, proving that habitable land extended further out to the east than at present. At West Tilbury a Romano-British hut circle was found buried on the Thames foreshore about 13 feet below high water mark at Clacton. In the last 2,000 years the south-east of England has twice subsided, and this has naturally affected the low-lying areas to inundation. In the second century AD there was the first dramatic subsidence in the Thames Valley; the evidence of Roman sites has confirmed that the land surface is now 15 feet lower than it was during the early days of the occupation. Dutch engineers and dike builders believe that the rise in sea level was not slow and progessive, but occurred in short and destructive bursts at about 200 year intervals. We learn, for instance, that in AD 245 thousands of acres of Lincolnshire were flooded by the tide and have never been recovered, while in AD 419 the coast of Hampshire was overwhelmed.

Those half-civilized communities living on the British Isles, like the Cim-

Devastating sea floods exacted a high death toll along the sinking coastlines of medieval Europe. This flood took place in Holland in 1430.

brians and the Teutons, were early victims of the rising tides of the North Sea as the sea filled with ice melt. Indeed, the relationship of the sea to British land was so significantly different from today that ancient ships used to navigate sea channels that have since dwindled or entirely vanished. For instance, Viking ships on the south side of the Moray Firth in Scotland were able to travel along a sheltered river that joined Lossiemouth to Burghead Bay. Since the fifteenth century the south-east of England is known to have been sinking again by about one foot per hundred years, thus justifying the building of the mammoth Thames Barrier at Woolwich. And London, built on a bed of clay, is even sinking faster than the surrounding country, which in some parts is rising.

Owing to the sharing of the English Channel coastlines, the fate of the British and her continental neighbours have been bound together. The west coast of France has shared in the disastrous sea encroachment of postglacial times. France is said to be sinking by about 30 centimetres a century. This seems to support the belief that the North Sea is a very recent basin. That tenuous land-strip between Britain and the Continent that vanished in late Subboreal times proves that seafloor spreading was made worse by vertical slippage as the swelling Atlantic bulldozed its way through the Straits of Dover towards Norwegian shores.

Vicious storms blowing in from the sea also added to the dangers to human life. Fierce gales on 11 November, 1099 struck English and Dutch shores with such violence that 100,000 people were said to have died. There was very heavy rainfall in the North Sea area from the twelfth to fourteenth centuries. The diarists and chroniclers of yesteryear have left written evidence to show that much loss of life occurred in 1100, 1134 and 1162. The sixteenth-century chronicler, Holinshead, mentions that in 1251 coastal seas rose by as much as six feet.

As the storms and hurricanes continued a Dutch sea wall collapsed in December 1287, and the Zuider Zee took some 50,000 lives. Again, on 18 November, 1421, the sea broke through the dikes, flooding 72 villages and drowning an estimated 10,000 people. There was, by all accounts, a terrible storm in 1446 with a further loss of life of '100,000' (round figures were always used in those days, probably to suggest a very high, but approximate, death toll).

The fate of old Dunwich, on the Suffolk coast, has been particularly well described by historians, and provides us with a vivid example of how destructive the elements had become in medieval times. Dunwich grew to fame after the Norman conquest, and prospered for 300 years. It had a fine harbour housing a large fleet of merchant ships, and was also an important farming centre. However, from the late thirteenth century onwards Dunwich came under ferocious attack from the sea and the sky. In 1287 a violent electrical storm damaged several churches beyond repair, inundated the land and swept away many buildings.

This was to be merely the beginning of a centuries-long nightmare, for as time passed Dunwich was to be battered literally out of existence. In 1328 a great storm filled the harbour with sand and silt. Little by little, as the sea pushed into the town, whole streets and houses vanished. The death knell sounded in the sixteenth century – literally with the clanging of chapel bells. In 1540 the church of St John Baptist was demolished, and before 1600 the chapels of St Anthony, St Francis and St Katherine were swallowed up.

But by this time the whole of Northern Europe was in chaos. At the time Dunwich was being pounded, sea floods on the Continent were drowning people in their thousands. Some 400,000 alone were reported to have died in a 1570 disaster. In 1634 there were 'great losses of land from the Danish and German coasts'.

Dunwich in the meantime saw its town hall surrounded by water. In 1715 the jail was absorbed. The climax was reached in 1740 when a north-east wind tore away the cliffs, destroyed the last remnants of the churches, and left nothing but gaping walls to mark the sites of ancient buildings. The cemetary of the previously destroyed St Francis chapel was burst open, and the few surviving horrified citizens of Dunwich could see the bones of the dead, some in complete skeletal form, being washed over the beach and along nearby streets. The Royal Castle of Dunwich, the seat of government of King Sigebert, its ancient glories, its thriving culture, were all swept away.*

Death-dealing Tidal Waves

The blame for much of the world's geophysical disasters can be attributed to outer space, or rather other celestial objects in space. Earth can never escape

---

* In the nineteenth century the town was competely rebuilt.

The crippling 'surge' tides that rolled down from the North Sea in January–February 1953 were the worst this century. A 1,000-mile stretch of Britain's east coast was overwhelmed, and nearly 2,000 drowned in Holland. This picture shows the aftermath at Canvey Island, Essex.

A rescuer calls out to a marooned house at Canvey Island during the 1953 floods.

Parts of Canvey Island begin to resemble Venice in this picture. Scientists predict that this could be a familiar sight within 200 years as ocean levels continue to rise.

Coastal towns like Mablethorpe in Lincolnshire are always vulnerable to sea surges and rising tides. As sea flooding increases, many inhabitants will be forced to move further inland.

60

from the baleful influence of its own moon and sun, and sometimes even the sister planets in the solar system.

Take the moon's influence on the behaviour of the oceans. This happens because the water on the side of the earth facing it is nearly 4,000 miles nearer the moon than the earth's centre. It is therefore attracted more strongly. Being liquid it yields in a highly exaggerated way (compared with the solid lithosphere held together by strong molecular forces), and becomes heaped up on the moon-facing side. This is known, of course, as the high tide.

The moon moves in its orbit in the same direction as the earth rotates, and the tidal bulge moves with it. And as the earth turns in its orbit, the continents pass through the higher bulge of water facing the moon. The waters ebb and flow with the high tide and low tide. On the side of the earth facing away from the moon the turning continents pass through the other tidal bulge about 12½ hours later, giving two high and two low tides every 24 hours. The sun, too, has a smaller part to play in the tidal phenomenon. The earth moves round the sun carrying its fluid envelope with it, but on one side the water is 4,000 miles nearer the sun than is the centre of the earth. It is this water travelling rather more than the earth's speed, that mounts up and gives a mini high tide.

All these tidal movements are ceaselessly automatic, and should not normally be a potential flood danger. However, twice a month the tide-generating forces of the moon and sun reinforce each other because they, together with the earth, are very nearly in a straight line. This reinforcement means that the tides are much greater than average, and are known as Spring tides, and can run up to six feet higher than normal. The sun, moon and earth also counteract each other over a similar period to create smaller tides known as neap tides.

If the earth were literally a Planet Acqua with no continental land masses breaking the surface, these tidal currents would hardly be noticeable. But the sizeable 30 per cent of the world that is solid succeeds in breaking the path of these currents, often with highly damaging results. The Pacific tides continually crash into Asia and Australia, while the Atlantic tides collide with America. So the coastal areas of the world frequently become flooded, especially when certain topographical features ensure that the effect of the tidal currents are more severe than they would be elsewhere. The Gulf of Mexico, for example, has an awkward funnel shaped by the adjoining coasts of Louisiana and Mississippi. These features make the tidal wave into a *tidal bore*, as the inhabitants of New Orleans and Gulfport know to their peril.

Britain in particular suffers from having a number of narrow funnels which can produce tides rising to 10 feet, such as at the Bristol Channel and the Severn Estuary, up which frequently rolls the Severn Bore. With the narrowing, constricting coastlines and shallow estuaries the water becomes sandwiched so that the height of the water builds up. Tidal bores reaching 15 feet have been reported for the Amazon, and even greater heights have been described for the Tsientse Kiang in northern China – occasionally as high as 25 feet.

Sometimes this bore is known as a *surge* – a fearsome term familiar to most Londoners, and which the new Woolwich Barrier is designed to keep out. The south-east of England and the coastline of the Low Countries in effect act as a

Artist's drawing of a flood in Monmouthshire in 1607. This catastrophic flood was probably the result of a massive sea surge emanating from the Atlantic that created a tidal bore along the river Severn.

The depth of these South Wales floods in early Victorian times were probably also the result of a tidal bore rolling up the Severn Estuary.

bottleneck. Hence when the tidal surge from the North Sea ploughs its way southwards towards the Thames estuary it gets jammed at the narrow opening at the Straits of Dover and the Flemish Bight. Then it 'surges' up the nearest opening – the estuary itself. The tide will continue to rush headlong up the mouth of the Thames until it becomes exhausted and the mechanics of the downstream flowing river take over.

And this is not all, since the surge is almost invariably accompanied by high wind speeds and localised reductions in atmospheric pressure which adds to the height of the water. The south-east of England and the Dutch and French coasts, still said to be sinking, are already very low-lying, with the ground surface several feet below Spring high tide level. So the effects when the surge happens are that much more severe.

Throughout the nineteenth and twentieth centuries, in particular, there have been some very serious coastal flooding, the most memorable of which was the 1953 surge. In that year in Britain some 307 individuals died; 32,000 had to be evacuated at the height of the disaster. From Scotland to Kent the sea smashed and destroyed some 24,000 homes and ruined much industrial equipment. In all 206,262 acres, over 300 square miles, of valuable British land along the entire east coast – a line of 1,000 miles – were submerged under life-destroying salt water. In some places not a mile of sea-wall remained intact. In Lincolnshire, where one-sixth of Britain's potatoes are grown, 17,000 acres were ruined.

The death toll was much higher, however, in the Netherlands, partly because the winds were blowing directly onshore. The dikes along the Dutch coastline were breached in 100 places, to flood more than four million acres of lowland. By 1 February, 1953 nearly one-sixth of the country was under water, with flooding most severe in Zeeland, Schowen and the Maas and Scheldt deltas. Nearly 2,000 Dutch folk drowned. Agricultural losses were severe, with nearly 10 per cent of the arable land submerged and nearly 500 farms totally destroyed. Farm animals came off worst – over a quarter of a million of them perished.

Italy, like Britain, has an exceptionally long coastline, and has had a despairing history of tidal inroads into its coastal communities. In November 1951 torrential rains, combined with high tides along the seafront regions, produced flooding throughout the mouth of the Po River valley, causing the death of about 100 people, 30,000 cattle, and causing damage equal to one-quarter of the Italian annual budget.

And yet there are always certain Italian towns, often historic and prestigious, like Dunwich, that seem to suffer unjustly. Take Venice, which to this day remains a victim of the rampaging sea – in this case the Adriatic (ironically not really a tidal sea) – sandwiched and honed between the coasts of Yugoslavia and eastern Italy. The high tides that have washed into the lagoons and canals that surround this tourist mecca have become increasingly grave. In a city where even at the best of times most piazzas and parlours lie no higher than 30 inches above sea level, and with canals lapping against crumbling stuccoed walls, a desperate solution is now being sought to relieve *la serenisima's* high water crisis.

Since 1876, the year when detailed record-keeping commenced, high water generally occurred twice a year at most. By 1930 that average had become seven

Tourists are familiar with the waterways of Venice. But frequently throughout the autumn and winter the water levels become much higher as mini surges rush in from the Mediterranean. This picture was taken in 1966 when Venice suffered from a very serious flood.

times a year. In the 1950s the figure had increased to sixteen times a year; today it is twice that. The most serious crisis occurred in November 1966 – the month when the flooding river Arno devastated Florence (see Chapter 7).

The crisis was compounded with Venice's slow descent into the lagoon, and by the blocking up of a number of artesian wells by the local authorities. The fate of the city was still determined by a sadly neglected system of dikes more than 400 years old which linked the islands separating the lagoon from the Adriatic. So lower portions of the town, like St Mark's Square, are often awash throughout the year. But the tourists see little of the drama, as the high water threat comes mainly in November and December. Then fierce winds gust in from North Africa and create a mini surge, causing huge waves to batter a prominent reef that separates the lagoon of Venice from the sea.

Eventually the waves became so fierce that they broke through a section of dike at the island of Pellestrina. Hurricane force winds lifted the water nearly seven feet above the expected level, which poured into St Mark's Square and flooded the cathedral. For 48 hours life in Venice was brought to a halt. Things were made worse by burst oil drums that spread slicks over the floodwaters. It was this oil-scummed tide that ruined the interiors of 4,000 shops and their stocks, as well as a similar number of private residences and hotels.

A movable steel barrier designed to save Venice from the fury of the Adriatic tides was approved in December 1982. But, because of bureaucratic wrangling over the submission of tenders, the city will now have to wait years before engineers can be contracted and work started. Yet Venice simply cannot wait. Floods now arrive at a rate of almost 200 a year, and even the smallest of them erode the foundations and walls of buildings.

The Dreaded Tsunami

Picture yourself on a Mediterranean beach in late summer. Suddenly the sea draws back in an exaggerated ebb tide to reveal a vast expanse of beach. The temptation is to head for the sand to collect a few easy lobsters and crabs. But if you were wise you would run, fast, in the opposite direction. For within less than half and hour the sea will surge in at a tremendous speed, bringing destruction and death to the seafront.

According to Michel Caputo of the University of Rome, such tidal floods are not as rare as people think in so-called 'tideless' waters. By scouring historical records he has found the earthquake-prone Mediterranean region as a whole yields a destructive tidal wave every two or three years. Around Italy's coast alone there have been at least 110 tidal waves since AD 1000. In the past 300 years, he calculates, land-based earthquakes have triggered many death-dealing seismic waves. He cites the Messina earthquake of 1908, where 60 foot tidal waves swept away more than a dozen villages.

The correct name for this phenomenon is *tsunami*. It is a Japanese word, and originally meant any form of tidal wave (*tsu* = harbour; *nami* = wave), but has lately come to mean a sea quake or seismic wave. They are caused by vulcanism that disturbs the ocean floors and generates waves that travel outward from the disturbance at great speeds. They often reach enormous heights when they arrive in shallow lagoons near the land, frequently taking vacationers by surprise. They can also be caused by volcanic eruptions, and by submarine landslides, and even by nearby man-made explosions.

The world's worst tsunamis disaster, one in which we can be reasonably certain of the death toll, occurred on the shores of the Bay of Bengal in 1876, leaving nearly 200,000 dead. There was an extraordinary wave that reputedly struck Pitdaea, Greece, in 497 BC, and led to the speculation that the Biblical Flood may have been caused by a series of catastrophic earth tremors hurling the oceans into populous coastal regions. We have already mentioned the destruction of the Minoan culture in 1450 BC following the eruption of Thera.

Although the mechanics of tsunami are still imperfectly understood, it is clear that it is not merely underground rumblings that cause the tidal waves, but a violent vertical displacement of the seabed itself. Since water is not compressible, an entire column of water, from floor to surface, is set in motion to race away from the quake zone. In the open ocean the waves are no more than a few inches high, even though travelling at high speed. When they pass a ship on high seas, those on board probably don't even notice them. And this in spite of the waves travelling at their fastest.

The speed of the wave is equal to the square root of the product of the acceleration and the depth of the water. So the deeper the water, the faster the waves travel. Thus a tsunami may reach a mere 30 mph in 60 feet of water, but exceed 600 mph in 30,000 feet of water. And this is measurable against a normal sea wave speed of no more than 60 mph. Naturally enough, as the wave approaches shallower depths its energy becomes concentrated, and it towers into an ugly looking column of water which then breaks on the shore at heights of up to 100 feet.

Eventually the waves die out or may strike an unpopulated continental land mass. Before doing so, however, they can travel half way round the globe, sending shock waves through all the world's interconnecting oceans. It would not be unusual for a Chilean tsunamis to strike China some 24 hours later. In 1755 an earthquake in Lisbon sent a series of devastating waves some 20 feet high rolling across the islands of the West Indies. Some waves have been reported to rebound, to actually slosh back and forth across the Pacific for more than a week. Rhythmic oscillation of water is known as a *seiche*, produced by less potent seismic activity, and often affects an enclosed body of water such as a lake or bay, giving rise to waves no more than five feet high.

It is the unfortunate Japanese who suffer more than most from these types of natural hazard. On average a wave 25 feet high is recorded once every 15 years. In the past 350 years Japan has been struck by fifteen major life and property destroying tidal waves. Probably the worst Japanese tsunamis occurred in 1703 when 30,000 people are said to have died. Another serious seismic flood occurred on 15 June, 1896 when some 20,000 people were holding a Shinto festival on an east coast beach about 300 miles north of Tokyo. Some 93 miles to the east, at 7 p.m., a light tremor shook the ocean floor. Less than an hour later another, more violent, earthquake occurred, sending a 110 foot wall of foaming water into the coastal town of Kamaish, collapsing homes and drowning 72 per cent of the population. The wave broke over Sanriku Beach, killing 27,000 and injuring another 9,000. Some 11 hours and 5,000 miles later, the tidal gauges in San Francisco began to flicker.

Sanriku suffered the same fate in 1933, this time becoming the worst tsunamis disaster of the twentieth century. Again an earthquake centred in Tuscarora Deep inundated the beach, killing about 3,000, destroying 9,000 houses and sinking some 8,000 boats.

The Pacific area undoubtedly suffers greatly from tsunamis. In the deep ocean trenches of the ocean itself at least one undersea earthquake has occurred every year since 1800. And a major Pacific tsunami of terrifying dimensions can be expected once every 10 years. Hawaii is particularly vulnerable because of its central ocean location, and has experienced thirty-seven in the past 125 years.

Hawaii's worst natural disaster occurred on 1 April, 1946. At about 2 a.m. a 100 foot seismic wave generated by a massive quake tore down the Scotch Cap lighthouse, 70 miles away, killing the 5-man crew. Notching up speeds of 500 mph, the waves surged southward silently until they crashed into the shore with waves more than 50 feet high. The city of Hilo suffered most, but throughout the islands 173 people were killed and over 1,000 buildings severely damaged.

Similarly, the coastal population of Chile are also highly familiar with the menace of tsunamis. The 1960 Chilean earthquake and floods were probably the worst this century after six extinct volcanoes exploded into life and three new ones came into being. A 30 foot high tsunami travelling at 125 mph struck the town of Corral, completely destroying all buildings. The waves surged against the east coast of New Zealand and Australia, and carried away sheep dogs chained to their kennels. The floodwaves then rolled towards Japan and deluged Honshu and Hokkaido with 20 foot high tides. Beachside homes were smashed and 300 people engulfed as they slept in their beds.

The inhabitants of Californian coast cities are seldom spared the knock-on effects of Pacific tsunamis, and have suffered from seismic sea waves originating several hundred or thousands of miles away. In 1964 about $10 million worth of damage was caused to the Californian coast when an Alaskan earthquake shook the whole of the North American continent. In Crescent City alone, where 30 blocks were flooded, $7 million worth of damage was wreaked.

Some tsunamis are more directly related to tectonic earthquakes when the epicentre is beneath the land, but also near the coast. A tragic example was that which affected the Himalayan mountain area of Assam, near the borders of India and Tibet. The tremor recorded was the largest since instrumental records were made. From 15th August 1950 onward the aftershocks continued for two months as the mountains readjusted themselves. The subsequent floods affected an area of 70,000 square miles along the Brahmaputra River, leaving a trail of desolation.

Occasionally the elements rage in unison, and quakes and tsunamis, typhoons and fires compete with each other to wreak the maximum desolation. Tokyo, for example, suffered a powerful tremor in 1923. The shock was so gigantic that landslides occurred wherever the land sloped. One obliterated a village and turned the sea red-brown for miles around. The resulting tidal waves, exacerbated by hurricane conditions, bodily carried a train and its 200 passengers out to sea. Yokohama, Tokyo's port, was reduced to rubble within minutes. When fire swept the city people fled to Yokohama Park, which itself was flooded with burst water-mains. Others who jumped into a pond to escape the flames were boiled to death.

Another famous but unusual tsunami was associated with the eruption of Krakatoa in August 100 years ago. It resulted in waves up to 115 feet high sweeping over the coasts of West Java and southern Sumatra. It destroyed the small town of Merak in the Sunda Strait, killing more than 35,000. About a dozen waves sloshed across the Pacific and Indian Oceans at speeds of between 550 and 700 km/h, and rolled across the Indian Ocean around the Cape of Good Hope. Some 32 hours after the explosion even the English Channel became choppy after the seismic tide traversed the Atlantic. The west coast of Panama and San Francisco city also felt the impact of Krakatoa, some 11,000 miles away.

Chapter Six

STORMSHOCK!

On the morning of 10th November 1970 an atmospheric depression formed in the Bay of Bengal, southeast of Madras. It rapidly turned into a cyclonic storm with winds reaching up to 55 mph. In the meantime Professor Raffaele Bendandi, of the Geophysical Observatory in Faena, Italy, had announced the observation of four separate groups of spots that were moving across the face of the sun. The observatory had already, the previous month, forecast that intense solar activity would bring disaster, and 'excessive cosmic activity' was still in progress. And by 6 a.m., on November 11th, things were beginning to develop at an alarming speed.

Wind velocities had increased to a hurricane force of 60 to 75 mph, and the storm had now centred some 650 miles southeast of Chittagong, and was hurtling across Burma. As it headed for the large islands in the Bay the water began piling up higher and higher. Then at midnight on November 13th there was a hollow, rolling roar and an eerie, cold, luminous glow. Suddenly a huge tidal wave – with a great, glinting lip of foam – struck the island of Manapura, at the mouth of the Ganges. In that brief instant some 16,000 lives were extinguished – three-quarters of the population.

That was just the beginning of the worst cyclonic storm this century that was to ultimately kill nearly one million through drowning, fatal injuries, and starvation and to seriously affect a further 1,200,000. The major islands in the lower Meghna estuary of the Bay of Bengal, such as Hatiya and Bhola, suffered badly. Vast stretches of the coastline near Chittagong, as well as about 2,000 tiny coastal islands, were suddenly inundated by the lashing storm-surge as it tore through a labyrinth of channels of little islands that form the Ganges–Brahmaputra delta. The wind speeds topped 125 mph, whipping up bores to a height of 50 feet with a murderous fury.

The storm wiped the low-lying land clean of animals, people and houses – most of them flimsy structures of bamboo, wood and tin. More than 1.1 million acres of cultivated rice paddies, with their estimated 800,000 tons of grain, were obliterated, as were one million head of livestock.

An Indian cargo vessel of 5,000 tons sank with her crew of 49. One agency report said that half the 200,000 population of Hatiya Island had been simply blown into the sea. Other islands – including the largest, Bhola – were under deep water with only the tops of tall trees showing. Later 10,000 bodies were recovered from Bhola and buried in mass graves or burnt on pyres. Of the 5,000 houses originally on the island, just four remained intact. In one house only eight survived out of fifty, and about 80 per cent of these were small children. The Noakhali district, on the eastern tip of the mainland, was also badly hit, with about 20,000 reported dead. In one particular Noakhali village a dam was blown over, killing 800.

In all, 25 of the offshore islands no longer existed as such – either geographically, as they were under water, or in ecological terms, since most human, animal and vegetable life had come to an end. Nearly 3,000 square miles had become a vast lake, the perimeters of the floodwaters only visible from the air after hours of flying time had elapsed. Eighty per cent of the population had been killed.

The aftermath of the flood was unpleasant. The ravaged areas reeked from the odour of decomposed human bodies and animal carcasses, some even found wrapped round trees. The paddy fields on Hatiya Island and on shore were blackened with salt water which destroyed the crop that at the best of times was highly inadequate. Only a few hundred head of buffalo survived out of 20,000 cows, sheep, goats. The main fishing industry was destroyed, and fields stripped bare.

The human heart of East Pakistan had been torn out. This delta land supported 80 per cent of the 70 million living at that time. This made it the world's highest density area, with about 1,200 people to a square mile, and with five million directly exposed to the danger from the Bay of Bengal's tidal waves. In much of the area the land is very low and flat, with only a few levees for natural protection. Much of the human ecology was of a subsistence level, with structures made with bare hands. Fishing and farming practices had remained unchanged for possibly thousands of years. The people are still among the poorest in the world, with a per capita annual income of as little as 600 rupees (compared with the 1,250 rupees for the richer Pakistan), or about £100 at the official exchange rate.

The Fate of the Bengalis

But the area has had a long history of similar death-dealing hurricanes. These fierce circular storms – occasionally known as typhoons in the East – in the Bay of Bengal are the most destructive in the world. The funnellike shape of the Bay tends to concentrate and hone the force of the storm. In October 1737, 40 foot waves roared into the Bay and veered up the mouth of the Hooghly River near Calcutta, killing an estimated 300,000. Again, in 1876, a tropical cyclone blew up near the mouth of the Meghna River. Massive surges caused waters 20 feet above the normal to flood the Bay's islands, killing 100,000 immediately, and causing the death of a further 100,000 later from starvation.

Bangladesh's flooding problems seem to have marginally worsened in postwar years. In October 1960, 3,000 were killed in a 70 mph cyclone which hit the Ganges Delta, followed by its 'worst tidal wave in memory'. Some 80 per cent of huts in the path of the cyclone were torn up as it raged for six hours across the Chittagong, Barisal and Noakhali regions. North-east India was affected as floodwaters rushed through Lucknow, bringing normal life for 750,000 to a halt for weeks. General Azam Khan, the East Pakistan governor, declared, futilely, that people must 'give up' living in thatched homes made of bamboo and mats.

The last tidal wave before the one in 1970 occurred on 11th May 1965, and swept 15,000 people and 50,000 homes and a similar number of cattle out to sea. Even in July 1970 six villages in the Delta were washed away, and by mid-August the number affected by flooding had reached two million. Yet again in August 1974 further disastrous flooding left five million homeless.

Geography is not the only problem for the region. For long centuries Bengal has been the marginal land of Indian empires and has been ruled by remote authority. The paradox is that the hiving-off of the two Pakistans from India after Independence in 1947 exacerbated this remoteness from both India itself and the East from the richer West. Although new jute mills were built in East Pakistan to replace those ceded to Calcutta after partition, much of the economic surplus of the East's jute-based foreign trade was transferred to finance the West's industrialization under Ayub Khan's hegemony. Compared to the wealth of West Pakistan the cost of flood control measures in the East would have been small: estimated by the World Bank at about $800 million.

Resentment and bitterness grew in the aftermath of the 1970 floods. The capital of Pakistan, Karachi, was centred 2,000 miles westwards across India. Some political observers have said this extraordinary geopolitical phenomenon was bound to ferment unrest. Rescue and relief efforts were tardy and half-hearted. In the meantime thousands more Bengalis had died of starvation and injury, and thousands more fled to Calcutta as refugees. The despair and chaos bred by the cyclone soon turned to rebellion and ultimately to fratricidal civil war. Eventually the rebels triumphed and the new state of Bangladesh was born.

Why the Elements Rage

How can we account for the violence of the elements that can affect the destiny of nations?

The hurricane is the very epitome of the weather machine. It is the product of two highly fissionable meteorological ingredients – heat and moisture. The world suffers storms because, to put it at its simplest, the masses of air around the tropics and the polar regions are essentially in imbalance. If this were not so a steady climate would be maintained everywhere. But the earth rotates around the sun, bringing on warm and cold seasons. In addition, the distribution of land masses, valleys and mountains in the northern hemisphere is not the same as in the south where, in any event, there is less land.

If a belt of warm air near the surface of the land gets trapped beneath a ridge of colder air moving down from the Arctic, a sort of conflict arises. A belt of air of a consistent temperature is called a *front*, as if it was the front of an army. The warm air will try to rise above the cooler air, and there is a battle to push one or other belt backwards. The conflict creates a frontal cyclonic system in the form of knotted bundles of low pressure, so turbulence and rapid air movements ensue. The winds generated begin to spiral round the columns of low pressure until they reach the centre, and are then forced upward.

The hurricane often has its genesis in late summer over warm equatorial seas where the air is laden with humid vapour (the more violent tornado is largely a land-based phenomenon). In the early stages there is a vast area of languid weather which drifts towards the centre-piece of the action where the pressure is beginning to sink and heated air starts to rise. The most violent features of the cyclone then begin to manifest themselves, slowly at first, but soon building up to a powerful climax when energy equivalent to an earth-shattering nuclear explosion is released. A normal cyclone develops at least a hundred times the energy of an atom bomb, and in the case of a much deeper pressure system the energy can be multiplied by a factor of ten. During the typhoon of September 1954, which destroyed over 300 homes and flooded 6,000 others in southern Japan, an official in the Tokyo Meteorological Office estimated the storm had the power of 'one thousand hydrogen bombs'.

To achieve this tremendous power the cyclonic system primes itself, like a giant toy spinning-top. The system begins to revolve, and soon whirls like a frenetic celestial body. In the southern hemisphere the winds rotate clockwise around the centre, at speeds approaching 150 mph, sweeping round in an arc and gaining momentum as they head towards cooler regions.

Hurricanes are a coastal phenomenon, since as they travel away from the sea they cut themselves off from the source of moisture, and then encounter friction from land-based objects. In a sense they are triggered into action by the contours and different temperatures of the land below. In the process they dump millions of tons of torrential rains, and an average US hurricane can yield as much as 20 inches.

The world's much used storm paths originate from the Pacific, and travel eastward along the Canadian border, to plunge southwards as the storm approaches the Great Lakes before spiralling through the St Lawrence Valley. Other North American hurricanes are deflected to the south towards the Gulf of Mexico, a familiar storm catchment area. In the autumn of 1944 a tropical hurricane swerved inland over the Mexican tableland and discharged millions of tons of water within 24 hours. Rivers flowing into the Caribbean from deforested watersheds became raging torrents, and more than 300 drowned.

But the worst US hurricane in history struck Galveston Island in September 1900, leaving 6,000 dead and causing over $30 million worth of damage. On 9th September President McKinley directed the War Department to 'hasten the supply of 50,000 rations and 1,000 tents'. Shortly afterwards, with the hurricane raging on its way to Newfoundland, social order had broken down as people died of hunger and thirst. Many bodies were taken to sea on barges to protect survivors from disease, as there was no hope of burial until the floods subsided. But this task was undertaken voluntarily, and men with armed Winchester rifles compelled others to load putrifying bodies on drays, and then haul them to quayside barges. Twenty-five were shot dead for allegedly looting valuables from bodies, martial law was declared, and Galveston was ultimately abandoned. It was many years before it was finally rebuilt.

71

The Menace of Hurricane Fifi

One of the worst of the Caribbean cyclones this century blew up in September 1974, and was again to have a profound impact on a nation's politics. Named Hurricane Fifi, it had been building up momentum for nearly a week. It finally swung westwards across the Caribbean to strike Honduras on Wednesday, 18th September, with torrential rains and winds over 110 mph, wiping out villages, small towns, roads and 75 per cent of the country's banana plantations. But, as with the 1970 Bengal cyclone, much of the damage and loss of life was due to flooding, with more than 9,000 dead and at least 100,000 made homeless by the time the storm had spent itself.

At one stage three-quarters of the land in the north of Honduras was under water. Choloma, 185 miles north of the capital Tegucigalpa, was the hardest hit, but might easily have survived. On the Thursday, an initial avalanche of water, earth and boulders crashed into the town but was held back by an earth dam. It was only when this burst that the town was deluged and the floodwater swept like an angry serpent through the stone and thatched houses. Here about half the

One of the worst cyclones this century struck Honduras in September 1974. Most of the damage was caused by the resulting floods, which drowned 9,000 and left 600,000 homeless. Here a house is washed up on top of a bridge in the Bajo Uguan region.

72

population of 6,000 died from the direct effect of flooding rather than violent winds that caused so much destruction elsewhere, such as in San Pedro Sula and the town of Omoa. In Cruz Laguna every single home had been washed away, and not one of the 1,500 inhabitants could be found.

Later, dark clouds of smoke could be seen rising from piles of bodies being incinerated in huge funeral pyres. Frightened and hungry families were reported to be clinging to trees and roofs in the north-west, while thousands were marooned on dykes in the Bajo Uguan region. Rescue efforts were hindered by washed-out roads, railway bridges and port facilities. From San Pedro, a town cf 85,000 inhabitants at the extreme south of the hardest hit region, it was impossible for rescue workers to penetrate into some 182 destroyed villages.

When the hurricane was finally spent, at least 600,000 people, mostly peasant farmers, were left homeless and without food. Water and slime 20 feet deep covered more than 7,000 square miles sandwiched between the two major valleys of the Ulua and the Aguan, which both overflowed.

Of all the seven countries in Central America that Hurricane Fifi might have picked, Honduras was the least able to bear a catastrophe of such dimensions. The country had its economy shattered. With a population of three million at the time, its inhabitants were the region's poorest, with a short life expectancy and subsistence wages of £80 a year. A classic 'banana republic', Honduras used to be known as the land of 'the Three Nineties' – 90 per cent illiteracy, 90 per cent infant mortality and 90 per cent illegitimacy.

In extent it is slightly smaller than England, but it has a fiercer topography – deep valleys, high mountains – and the hand-maiden of flooding disasters – fast-flowing rivers. The country has three seasons: the wet from May to September, the dry from November to January, and then the hurricane season for the rest of the year.

In 1870 vast sums of money were borrowed from London to build railways. But the investment funds were squandered through incompetence and corruption. The country was soon bankrupt and ripe for revolution, and from 1883 until 1944 the people experienced little else. Then, in 1969, the leadership of the country decided to withdraw from the newly formed Central American common market but succeeded only in dealing a severe economic blow to the already impoverished food producers of Honduras. The only industrial growth area – around San Pedro Sula – was soon brought to ruin years before Fifi finished off what was left. And then, in the late 70s, wars broke out with other Central American neighbours, and terrorist death-squads stalked the land. Now the only memory westerners have of Honduras is one of civil war, insurgency and East–West power politics.

The message of the Bengal cyclone of 1970 and Hurricane Fifi of 1974 was clear: little can be done to aid Third World victims of natural disasters. In both affected regions some were warned of the impending disaster but most were not. According to one press report, loss of life in East Pakistan on such a stupendous scale could have been averted but for a sudden decision by Pakistan's Meteorological Department to discontinue the usual system of announcing storm warning signals on the radio. People were intrigued on the night of the disaster to hear radio warnings about the cyclone without mention of danger signals. In any event the vast majority of the populations concerned had nowhere to flee to. There were no safe zones, no refugee camps, little official help, rescue or aid. The people had to remain where they were and await their fate.

This became clear when cyclonic storms returned to Bangladesh at the end of May 1985. The meteorological authorities had seen the depression forming several days earlier. In spite of the reinstatement of the cyclone warning system, the death toll was still exceptionally high. Indeed, the pendulum had probably swung too far the other way, as the broadcast warnings had been excessively strident since 1970. In most cases the depression either dwindled or veered away to some other adjacent coast. Those few inhabitants who had the option of moving out of the area preferred to take the risk of just staying put.

As a result an estimated 10,000 were killed. After heavy and continuous rain brought severe flooding to that country's eastern delta, a giant tidal wave and cyclone hit six main islands and the coastal belt. A quarter of a million families were affected. Nearly 17,000 homes, many of them single-roomed bamboo and thatched huts, were demolished, and 122,000 severely damaged. Over 140,000 head of cattle were lost in an already malnourished nation, and nearly 500,000 acres of rice and jute were ruined.

Heavy rain continued for days afterwards in the districts of Sylhet and Comilla, and on the banks of the Gunti River in Uttar Pradesh. Some 250,000 were forced to move to safety when rivers rose five feet above danger level. Worst hit was Urirchar Island, where more than 6,000 lived, which was wiped clean of human habitation. Many of the bodies had been swept out to sea; there was no final body count.

The 1985 cyclone, however, highlighted more than ever the pernicious role that over-population plays in human disasters of this kind. Bangladesh is simply the most populous nation on earth. Urirchar, which for years earlier was merely a sandbank, shows the extent to which the inhabitants will colonize even a reclaimed flat stretch of dried laterite mud. Because it was frequently submerged it could be made to produce a higher than average crop yield. But the price of living so close to the sea is a high one.

The Mechanics of the Deluge

Let us now look for a moment at the main causative agent of floods in other parts of the world: relentless, and often torrential, rain.

We have already seen in the introduction that the air is loaded with moisture. But what causes the moisture to precipitate is the convection of residual surface heat which tends to push the air upwards until, more often than not, it reaches a highter belt of cold air. Arctic winds, seasonal variations, etc. all play a part in this cooling process, but we must bear in mind that it is one of the characteristics of warm air to suffer a cooling fate. For the further the air rises above the land, the less the atmospheric pressure, and this tends to cool the air. If the air is relatively full of vapour quite a small push upwards will cool it sufficiently to squeeze out the water droplets, and low cloud will appear. The less moisture in the air, the more it will need to ascend before a rain-cloud will form, and the colder the air at the beginning of its ascent, the less moisture it is able to hold.

As the tiny drops inside the cloud continue to coalesce, forming larger drops, the weight causes them to fall as rain. This is a regular feature of life that makes 4,000 million tons of water fall upon the globe every year, which averages out at 40 inches for all the world's land masses.

As it is the rising, cooling air that causes rain, rain-clouds are more likely to form over areas where there are mountain ranges which tend to cool off the rising warmth of the land as it creeps up the mountain slope, and least likely over larger flat areas. On the Peruvian coast, in the rain shadow of the Andes, the inhabitants would be lucky to receive two-hundred-fiftieths of an inch; a man could easily tramp through coastal valleys sheltered from the wind and come across his own footprints made 20 or 25 years earlier. And in polar and subpolar regions the air is so cold it can hardly retain any water vapour. Conversely, the areas of highest rainfall are at the equator, with over 60 inches a year.

It is at this stage that the 'heat engine' takes over, and, as in so many earth processes, a self-balancing effect occurs. It is the sun's heat which causes surface evaporation and cloud formation, but the clouds in turn will prevent the sun from warming the land below. According to the German climatologist Hermann Flohn, the skies above the Sahara Desert contain as much water vapour as those over the fertile lands of Northern Europe. But the reason the Sahara recieves hardly any rain at all from one year to the next is because the sun burns so fiercely it can actually vapourize any tiny clouds that do appear. Deserts have a high rate of reflectivity, too, unlike the northern hemisphere with its patchwork of urbanized, wooded and arable surfaces. And by definition, of course, deserts have no surface water to evaporate into clouds.

We should not neglect the role of the seas, which act as great heat reservoirs. They absorb much of the carbon dioxide emitted by dying vegetation and Man's burning of fossil fuels, so preventing terrestrial temperatures from rising excessively. The seas also moderate the high land temperatures of summer and the cold of winter. The oceans cause what is known as the general circulation. They generate thermal instability in the land/ocean heat ratio, and push the warmer equatorial waters up to the polar regions and down again.

Some countries benefit considerably from the fact that they are surrounded by water. The Gulf Stream carries warm water eastwards across the Atlantic to make the British Isles much milder than their latitude would warrant. On the other hand Britain is a wetter country, on the whole, than its European counterparts. Being surrounded by the heat reservoirs of the seas, the atmosphere becomes heavily burdened with moisture. So it becomes harder for the sun to vaporize away clouds as often as it does on the Continent.

The disadvantages of being surrounded by seas is made more obvious in the tropical regions. In south and east Asia heavy, moist-laden air gives rise to the *monsoons*. The summer low pressure centre which originates over the northern plains of India and Pakistan extend deep into the interior of Asia, drawing in moist, warm air from both the Indian Ocean and the western Pacific. There is a dramatic reversal of the winds when the winter high pressure system gives way to the summer low, bringing on storms and torrential rains. It is the associated hurricanes and monsoons which make the Indian sub-continent continually prone to floods, with annual losses of 700 people, 40,000 cattle and £100 million at 1983 prices.

Record-breaking Rain

The 'Weather Machine', as Nigel Calder reminds us in his book of that name, has not the precision of a real machine. Indeed, the un-machine-like properties of the world's weather should be obvious from a glance at a cloudy sky. Clouds vary enormously in extent, size and density. Rainstorms covering more localised regions are usually intense, being the product of dark, woolly shaped cumulo-nimbus clouds, and are usually finished within a couple of hours. The thinner stratus cloud produces less rain than the former, but as it covers more of the sky it often rains for longer periods.

The intense type of storm is often known as the cloudburst or 'gullybuster', and was once suffered by the town of Louth, Lincolnshire, in 1920, drowning twenty-one people. The village of Langtoft, Yorkshire, was similarly over-whelmed by an intense cloudburst in July 1892. In this storm, two cottages were demolished, some hill fields were washed away to the bare rock, and other houses were left with a two-foot carpet of mud.

Many of these cloudbursts can result in several inches of destructive rain coming down within a short space of time. The layman, however, especially an Englishman used to frequent downpours, will often express surprise that a 'few' inches of rain can actually cause damage to property and loss of life. The picture can be clarified, firstly, by getting an idea of average global rainfall figures. Among the world's major cities an annual average of 1.2 inches falls on Cairo, and nine inches on Singapore. London and west European capitals receive 24 inches, and New York some 43 inches. It is easier, with these statistics in mind, to get a better idea of what must happen when average rainfall measures are improved upon many times (sometimes hundreds of times) over. If New York averages 43 inches a year, what are we to make of the record 28.7 inches received during a 24-hour period during a hurricane at Yankeetown, Florida, in early September 1950?

Then there is the important concept of temperature/precipitation ranges. Boston receives nearly twice as much rain annually as London (normally considered a damp city) because her July temperatures are about 8°F warmer. Hence the city loses much more summer moisture to evaporation, and is therefore more prone to cloudbursts.

Record rain statistics are related to a time factor, so that inches of rain are measured against hours, minutes and seconds. Each time-segment seems to have its own record listing chalked up in meteorological bureaux. The United States, with its volatile and often neurotic weather patterns, has received some extraordinary record-breaking deluges this century when measured against a monthly average of two to three inches. Because of America's sub-tropical climate many downpours, of course, are storm induced.

Often, as we have seen, hurricanes are centred around the Gulf of Mexico. Hurricane Camille, which blew up in August 1969, broke a record because of the torrential rains it unleashed in the Upper James River basin of Virginia. On August 19th more than two feet of rain descended in eight hours, and drowned Lynchburg and Scottsville. A US Weather Bureau meteorologist described rainfall of that magnitude as occurring 'only once in more than 1,000 years'. As many as 107 Virginians lost their lives, with 55 missing and presumed dead. Another incredible downpour caused the July 1976 floods at the Big Thompson Canyon, at the edge of the Colorado Rockies. Within the space of five hours, 10 inches of rain fell just below Glen Comfort, a little community nestling alongside the canyon highway, US 34. The annual average rainfall for the area would normally be less than 16 inches.

There have been other notable American rainfall records. The maximum precipitation for five minutes is 2.03 inches at Alamorgordo Creek, New Mexico, on 5th June, 1960, and that for *one* minute was 1.23 inches at Unionville, Maryland, on 4th July, 1956. Seven inches of rain fell in just 30 minutes at Cambridge, Ohio, on 16th July, 1914.

Of growing importance in recent years is the 'flash flood', where an unexpected and torrential downpour falls on an impervious surface like a dried out river bed, or a hill terrace baked rock-hard by the sun. What aids the flash flood is a convenient dry channel. Even before Noah's Deluge, judging from the scattered and wretched remains of a small family group found in Afar during the summer of 1975, a flash flood is believed to have raced down a dry Ethiopian river bed some three million years ago. The worst flash flood in history allegedly killed 2,000 religious pilgrims gathered near Teheran when the storm waters tore through a normally arid valley.

Aqualanches – the liquid equivalent of an avalanche – still claim many victims every year, the most recent being in 1972, when 237 were killed in Rapid City in South Dakota. In another instance, in southern Spain in 1973, clay soils had been baked to the hardness of granite during a long heatwave, and heavy autumn rains were funneled along dry channels to take the lives of 150, and demolish houses that had stood safely in more moist times for over a century.

Even without the convenient gulley, flash floods can uproot the entire geology of a valley: earth, trees, rocks and vegetation are usually swept clean away. In July 1883 such a flash flood poured down a slope in Utah. Complete mounds of earth, with willows standing erect, came careering down with the torrent, and when the flood ended a canyon was some 50 feet deeper than it had been just eight hours earlier. Another notable flash flood struck Willow Creek, Oregon, in 1903 and wiped out one-third of the town of Heppner and killed 200 in less than an hour.

What the flash flood tells us, then, is that the area/intensity relationship, as well as quantity, is highly relevant in assessing whether excessive rain will cause floods. For example, a six-hour storm that deposited five inches of rain over a 50 square mile area might only deposit some 2.75 inches over 1,000 square miles. And if a 12-hour downpour covered only 10 square miles, we might expect about 35 per cent of the rain to fall within one hour. Yet over a 1,000 square mile area, less than 20 per cent would be expected to fall within one hour, and only 40 per cent in three hours.

Wind speeds can play an important role in rain-making, since they can actually prevent the rain from precipitating. The drops then accumulate in the cloud, which grows bigger and darker as if it was being packed with a lethal charge. Thus the deluge often starts when the wind speeds drop. When the air currents suddenly cease, a vast accumulation of rain hurtles down in a matter of minutes, whereas on a still day only moderate rainfall would more likely occur.

Tropical areas, needless to say, frequently experience dramatic downpours. Take, for instance, the unbelievable *480 inches* of annual rain that falls over the Khasi Hills in Assam on the north-east tip of India. In some years twice that amount falls. And from August 1860 to July 1861 nearly 90 feet of rain was totted up from scattered rain gauges. Another exceptionally rainy spot is Mount Waialiali, Hawaii, with annual precipitation exceeding 400 inches. The world record for 24 hours of rain remains 73.62 inches on the island of La Reunion in the Indian Ocean on 15–16 March, 1952.

The destructiveness of Asiatic rains cannot be overestimated. In July 1982 the death toll in Nagasaki and the neighbouring areas of west Japan approached 300 when torrential rains caused massive landslips. It was the worst such disaster since 1957 when abnormally heavy rains set off floods in Isahaya, also in west Japan, killing over 600. This time as much as 24 inches of rain fell on Nagasaki in one day.

In some places – Singapore is a good example – rainfall is more or less evenly distributed throughout the year. In other areas, like Sicily, the rain is concentrated in one short spring season, and very little falls in summer; quite the reverse in Kano, Nigeria, where the summers are usually very wet. In other places, especially other parts of Africa, and Australia, the rainfall is highly irregular from year to year.

The most extraordinary rain flood of recent times happened in Latin America. In Brazil, in March–April 1974, nearly two weeks of torrential rain drowned about 20 per cent of the country across 22 states, from the dry north-east to the agricultural south. Over 1,500 people lost their lives.

Spells of prolonged rain of several days' duration are now usually associated with slow-moving depressions. The entire sky becomes grey and overcast, seemingly unchanging and unending, with almost stationary air currents. In spite of the absence of the traditional woolly cloud, the water-vapour content of the air is high, and precipitation is frequent. Whereas in the eastern hemisphere intense, cyclonic storms wreak widespread damage over quite short intervals, continuous rain is more often a problem of northern Europe. There was, for example, an unbroken downpour in London from 1 p.m. on 13th June to 11.30 p.m. on 15th June, 1903 – a total of 58½ hours of rain. There was another three-day deluge from 20th to 23rd July 1930, when nearly 12 inches of rain fell at Castleton, in the North Yorkshire moors, causing widespread flooding. Such rainfall figures would be microscopic in comparison with America and the Far East. But it must be remembered that London can usually expect no more than one inch for its average wettest day. In the hilly country of Cornwall, Devon and

Wales the wettest day probably yields two inches of rain, rising to three inches on higher ground.

Even so, Britain is still universally considered to be a rainy country. Although a scorching summer, as in 1976 or 1983, makes a welcome change for the long-suffering holidaymaker or tourist, even these hot spells are followed by record-breaking rainy spells, as in the former year, or preceded by them, as in the latter. In the first half of 1983, in fact, Britain had its coldest, wettest spring 'on record', and there was much talk of atmospheric disturbances caused by dust particles ejected into the air by the Mexican volcano, El Chichon. April and May 1983 experienced frequent downpours, with 40 consecutive days of rain. But this might simply have been bad luck, where one depression drifted away and another moved in to take its place.

Another theory, which we will leave to a later chapter, is that 'blocking' ridges of low (or high) pressure high in the North Sea or over Eastern Europe keep depressions stationary over the land for longer periods than usual. Britain often, in fact, gets sandwiched between ridges of low pressure coming both from the Atlantic and from the East.

Chapter Seven

EARTH'S RAMPAGING RIVERS

If man has suffered enough over the centuries from inclement cyclonic weather, it is worth considering the storm's disaster-prone handmaiden – the river. When the two agents of doom work in unison, an incomparable death-dealing mixture befalls mankind. Let us turn our attention to the disaster that afflicted Italy in the autumn of 1966, one of the saddest tragedies that that beautiful but troubled country has had to endure, and Europe's worst postwar river flood. Both Venice and its Renaissance cousin, Florence, were to suffer appallingly. The total desecration of an important part of our western cultural heritage was accomplished in just 24 hours: no wartime bomber attack could have inflicted so much damage.

It began, as so often, with a severe atmospheric depression that plagued the whole of continental Europe, reaching the shores of the North Sea, and then racing round the Skaw and ripping into the Baltic. Then it reached the Mediterranean after roaring through France and Switzerland to finish up in Yugoslavia. There was the inevitable death and mayhem, but it was only when the gales and rain hit the mountain streams and rivers of Italy that the storm's destructive potential was fully realized.

Italy's rivers, meandering down from high mountain ranges, experienced six months' average rainfall in 24 hours, and quickly overflowed, bringing down with them masses of rocks and debris from the Dolomites and the Alps. The pulverizing force of the fast-moving waters was so great that vast expanses of riverside forest were reduced to so many wood chippings.

The River Arno was the main agent of disaster. Throughout November 4th the waters of the Arno rose, until only three feet of space was left beneath the Ponte Vecchio. Enormous stretches of road surface were completely washed away, bridges were reduced to rubble and railway lines twisted grotesquely. Other tributaries were so full of broken trees, boulders and rubble that they were many feet higher than the roads running alongside. The waters surged through Florence at 40 mph, flooding the city's storm sewers, and inundating cellars in the districts of San Croce and San Frediano. Many of the Ponte Vecchio artisans and shopkeepers had been forewarned by watchmen, and arived at their shops with barely enough time to rescue their priceless stocks of gems and precious metals.

By 7 a.m. all electric power had failed in the city, which was both cut in two and cut off from the rest of the country. Three hours later floodwaters seeped into the Piazza de Duomo and poured into basements. Oil storage drums for central heating burst open and a sticky sludge-like veneer was added to the half a million tons of mud, to leave its indelible mark on walls. The foaming floodtide, honed by narrow streets, now reached speeds of up to 60 mph, and washed into some 5,000 homes. Barely half of Florence's economic infrastructure survived.

Meanwhile in Venezia-Giulia, in the north-eastern province reaching down to the Adriatic, scores of farm animals were overwhelmed by aqualanches, landslips and mudslips. In Treton, Merano and Bressanone streets were so deep in mud that entire cars and buses were swallowed up.

The following day the Interior Minister, Sr Taviani, described the floods as 'the worst in history'. In the Po Valley the damage from the overspilling river was compounded by the tumultuous sea which broke through dikes protecting the delta, one-third of the population in the region having to be evacuated. Sadly,

Florence was all but submerged in November 1966 when the River Arno overflowed. This was Europe's worst post-war river flood.

the thriving industrial region of Tuscany was decimated, barely 20 years after it had been reclaimed from its malaria-ridden marshland state. Four-fifths of the town of Grosetto was submerged, and 80 per cent of its livestock perished.

Florentines were later shocked to learn that literally millions of rare books, 730,000 valuable ancient letters and manuscripts, and 1,300 paintings and etchings were damaged or destroyed. In the Biblioteca Nazionale 1,300,000 volumes were damaged; at the Library of the Jewish Synagogue some 14,000 books were ruined; at the Gabinello Vieusseux a quarter of a million were lost, as were 36,000 at the Geography Academy. At the Music Conservatory the entire collection was demolished. The damage to Florence was heartbreaking. The thirteenth-century church of Santa Croce would have to be completely rebuilt, as would the San Firenzi Palace. The frescoes in the Medici Chapel began to blister and peel from the walls, while 600 paintings by well-known masters were under water for hours in the famous Uffizi Gallery. Buonarroti House, on the banks of the Arno, famed for its Michelangelo drawings, was completely flooded out. The illuminated manuscripts in the Cathedral Museum were irreparably ruined. Elsewhere the entire State Records of Tuscany from the fourteenth to the nineteenth century were lost, as well as the Etruscan collections in the Archaeological Museum.

After the deluge, some 24 hours later, 17 persons were left dead in Florence, and 4,000 rendered homeless. As many as 6,000 shops were ruined, including the workshops of some of the leather workers and goldsmiths that the city was renowned for. Throughout Italy some 800 municipalities had been affected by the storm of that November, and 22,000 farms and private homes suffered. There were 50,000 amimals lost and much valuable farm equipment made useless. The total Italian death toll was 112.

This was not the first devastating Arno flood. The Arno Valley was first inundated in 1117, sweeping away the original Ponte Vecchio for the first time. Florence's city walls collapsed in 1333, when four feet of water surged through the town centre and knocked down four bridges, drowning some 300 people. Since then, before the 1966 disaster, Florence has suffered the misfortune of over fifty other serious floods.

When Rivers Can Take No More

What happened at Florence was a salutary reminder of the cruelty of earth's geography. Without rivers, torrential rainstorms would seldom wreak the kind of damage they have. Even small streams can be devastating transporters of floodwaters. Many of the heavy bouts of rain written about in the previous chapter were collected by watercourses of various sizes and dumped on to the hapless residents of the river valley concerned.

Italy represents a case where this geographic fact of life is writ larger than usual. Many parts of the country have been subject to heavy flooding because Italian rivers rise thousands of miles above the plain in the Alps, the Dolomites in the north-east, and the Apennines running down the centre of the peninsula. Hence, unlike most of the other major European rivers – the Seine, Rhine,

Thames – they have farther to fall, and so can be lethal.

Many early geographers tended to write about river floods in rather circular language. A flood is simply an accumulation of water that cannot be absorbed into the soil; an overflowing river occurred when 'the channel capacity is exceeded by the runoff from its catchment area', and so on. The task then was to list certain river characteristics that contribute to overflowing: log or ice jams, snowmelt, sudden changes of direction, broken dams, siltation of the stream bed, a reduced gradient causing the flow to pile up, and so forth.

Some critics have pointed to the lack of predictive power among geographers and hydrologists, and have cited the officials at upstream levels of the Arno on the fateful morning of November 4th who observed speeds of up to 36 mph, but failed to act.

In practise, however, the situation is so complex as to preclude all but the most rudimentary predictions. There is no real coefficient between rainfall and river levels. This is testified to by local riverside inhabitants who have stood vigil many a time while expecting a drowning that never materialised, and by those instead who have been flooded out after only moderate falls of rain. A frequently unsuspected variable is high winds that can greatly exacerbate the dangers. Winds can actually hold back floodwaters for a while, only to release them later.

In addition, no two rivers in the world have the same run-off absorption characteristics, and each river delta has its own contours and subsoil qualities. There is the vitally important – and often neglected – fact of *infiltration*, meaning how much water is absorbed into various types of surface as a general principle and for specific areas in particular. Some river plains have rich vegetative cover. This is an important flood alleviator, as it improves the soil structure. And vegetative roots act as channels to divert excess water into the ground to greater depths.

The high humus content of grassy flood plains acts rather like a sponge. Soils with large pores or fissures, such as sandy grit, or gravel, allow the maximum infiltration (or percolation), while clay soils are the least permeable in ther usual moist (as opposed to dry and cracked), state. The underlying geology is important too, as a second more impervious layer of rocks can counterbalance a more porous surface. And rain may also fall only on one small part of the delta, or on the leeward side away from a slope.

The question of infiltration is related (not precisely, of course), to the amount of rainfall, with or without the intermediary of rivers. But the historic shape of river valleys has often been determined by earlier run-offs, as the flow has channelled its way along gulleys and depressions. So those valleys with hard layers of stone or shale on the surface will tend to expand sideways rather than downwards.

And we must not neglect the phenomenon of the *meander*; the way the track of the river has wandered throughout time and carved a zig-zag course for itself over the countryside. While the river nibbles away at the outside bends it deposits material on the inside of the channel. Then, with its ability to build and enlarge its own channel – to create its own flood hazards – comes one other important feature. And that is the levee. This is the name given to the natural mud wall or embankment on the lower Mississippi or the swift muddy rivers like China's Hwang Ho, but it applies to rivers in general.

The overflowing river deposits its sediment against the banks, continually raising the height of the stream channel above its surroundings. Levees are often strengthened artificially, but during a severe flood they are easily burst through. And because the flood will then fall from a higher elevation, the catastrophe is that much greater. Some of the worst floods on the Mississippi and Hwang Ho have occurred in this way.

Rivers are hence geographical tautologies. They originate as subterranean streams but largely create their own watercourses from sediment transported by themselves. Then the river goes through a period of adjustment to its self-modified flow and to the volume and nature of the sediment. But this adjustment is invariably upwards and outwards: the waterway grows continually providing the flow is continuous. Even in its drier phases it remains a potential threat to life and property because of this ratchet effect. Like a self-fulfilling prophecy the river in its floodstage further increases its power to erode the channel and valley sides, thus making the next overflow more disastrous than the one before. And the potential for damage is the extraordinary power of the flooding river to shift boulders further downstream to occasionally force the waterway into a totally new route towards inhabited areas.

This brings us to the question of velocity. For most rivers passing through a great drainage basin the average speed of flow is much less than one mph, but at the flood stage it can reach up to 20 mph. Hence those serious floods in England, where normally the river plains are lower, naturally occur amongst the most steeply sloping rivers. In the case of the East and West Lyn, in the hilliest part of Devon, the gradient is almost a precipitous drop. The catchment areas of both rivers consist of steep, narrow ravines cut from the hard slates and grits of Exmoor.

Lynmouth is a picturesque seaside holiday resort tucked beneath the thickly wooded hillside, and, as its name implies, it is situated right at the delta of these steeply tumbling torrents. There had already been disastrous floods in 1607 and 1769, and it was statistically highly probable that a major twentieth-century flood would occur. This probability came to fruition in August 1952. The ground was already wet from two weeks of rain, and on the fifteenth of the month a fierce storm over Exmoor yielded over nine inches of rain. This was a phenomenal amount by British standards. The West Lyn became a bursting murky torrent, rising fast. Because of the steep gradient the velocity achieved was quite extraordinary, probably faster than the Arno in 1966, and dislodged great chunks of earth and boulders. This added to the battering force of the onslaught which smashed into all obstacles in its path. Between them the two rivers poured millions of gallons of water down gorges into the town below.

Sewers and electric cables were ploughed up or put out of action. Buildings were flooded to five feet in just three minutes. The turbulent waters carried all before them, undermining 23 buildings, demolishing 29 bridges, and causing 38 cars to disappear out to sea. Twenty-four victims met their deaths, and 400 were made homeless. Half a mile out to sea hundreds of trees, presumably weighed

These pictures illustrate the enormous destructive power of river floodwaters when they fall from a great elevation as they did here, at Lynmouth, Devon, in 1952. Entire buildings can be gutted, and the dislodged debris acts as a battering ram to weaken structures further downstream.

The aftermath of the 1952 Lynmouth floods.

down by rocks and soil entangled in roots, and with their upper branches above the waves, looked from a distance like a fantastic sea forest.

Some observers, however, believe that the human factor played a big part in contributing to this serious flood. Because of the desire to preserve intact an attractive hillside beauty spot, the West and East Lyn rivers had not been widened to accommodate a greatly increased flow. It was calculated that the West Lyn bridge, whose arch had become blocked, was only half the width needed to facilitate the sudden downrush that August of some 10,000 million gallons.

Engineers and hydrologists insist that in any given year there is a 10 per cent chance of a river rising above its previous height. So in theory there could be a disastrous flood every 10 or 20 years, and a spectacular flood can be expected once a century. Most experts agree that more attention must be given to communications. In the Lynmouth disaster many were not able to take evasive action, to vacate their houses promptly because telegraph poles were torn down before phones could be used. And in 1979, when there was a growing flood threat in Devon, some of the phone lines sputtered out of action, and about a third of the rain gauges stopped working in the most critical areas.

The Yellow Peril

There has not been a single river flood in the world that can conceivably compare with the unbelievable death and devastation meted out by China's two great rivers – the Yangtse Kiang and the Hwang Ho. These two rivers have had a chequered, restless and violent career. The Hwang Ho especially – aptly known for centuries as 'China's Sorrow' – has a reputation for causing the deaths of more innocent human beings than any other agent of natural disaster anywhere in the world. Literally untold millions throughout history have been drowned in this mighty, swirling torrent. In the past it has claimed, and can still do so today, the lives of several thousand people during one overspill. And at intervals of about 150 years or so, commencing in 2297 BC, the death toll tops the million mark.

On its 2,600 mile meander from the border with Tibet it picks up enormous quantities of wind-blown yellow sediment consisting of fine particles of clay and dust, and carries it out to the Yellow Sea. Not surprisingly, this sepia coloured waterway is often known by its alternative name – the Yellow River.

For 4,000 years this ungovernable river has wandered over its wide river plain through the most densely populated parts of China in which up to 100 million people live. In some places the river is a mile wide, but its chief and dangerous characteristic is that it often runs well above the surrounding flat and fertile land. It is somewhat shorter than the Yangtse and, surprisingly, carries less water and less traffic. It does, however, greatly compensate by carrying vastly more silt.

After a circuitous eastern journey from the Tibetan Highlands of over 2,600 miles across China, the Hwang Ho arrives at the great delta of the Gulf of Po Hai. Some 500 miles before it reaches the delta it halts at Kaifeng, and then radiates into 15 channels across the plain. At one time the waterway coursed northwards by Tientsin, and at another time it has joined up with the Yangtse Kiang near Chekian, several hundred miles to the south. Today the Yellow River continues to pose an ever-present threat of death to the 50 million farmers and peasants ekeing out a subsistence living on the 50,000 square miles of farmland in the Yellow River Basin.

The reported death toll from the Yellow River strains the credulity of the serious researcher who tends to believe that the number of fatalities, like many distant historical events, may have become embellished with the passage of time. Professor Hubert Lamb, normally a highly reliable scholar, writes in his Volume II of *Climate: Past, Present and Future*, that Chinese river floods in 1332 accompanied by heavy rainfall took the lives of as many as seven million. This was undoubtedly a disastrous flood, since Lamb claims that it destroyed various sanitation arrangements, encouraged the spread of plague-carrying rodents, and probably brought about the Black Death in the following year.

The Yangtse Kiang has also taken its dreadful toll. The Yangtse is China's longest river, running, like the Hwang Ho, from Tibet, and flows for 3,400 miles, entering the sea just north of Shanghai. The plain and delta of the Yangtse is also heavily populated, and in fact is known as 'China's Main Street' because of the predominance of river traffic. It took the lives of 100,000 in 1911. In the great Yangtse flood of 1871 the waters were said to have reached a height of 275 feet above its normal level in the gorges downstream from Chunking. In one instance a river steamer was spotted marooned on a rock 120 feet above the river when the waters subsided. And in June 1931 another disastrous Yangtse flood drowned more than one million.

In July 1981 Sichuan Province, once called 'the grainbowl of China', faced its worst floods in more than 30 years after days of torrential rain, causing chronic rice and grain shortages. Troops and aircraft had been rushing relief supplies to the stricken areas where peasants, aware of the coming threat, had sat out the night on roofs. The Yangtse, swollen by four tributaries, had risen in places by as much as 16 feet. And in Guangdon Province some 250,000 acres of ricefields had been damaged by floods in a single prefecture. Twenty foot waves were generated when the Yangtse River rose to its highest level this century. Jiangjiang, an area of fertile rice and cotton fields in the central Hubei plain, was inundated. China's largest hydroelectric dam, the Gezhouba, withstood a surge of floodwaters pounding up to six metres high. More than 3,000 were killed in the province, and a similar disaster afflicted the north-eastern province of Shaanxi, where more than 700 were reported dead and 20,000 made homeless.

The damage caused in the Suchuan floods was estimated to be worth £700 million. It was particularly disturbing for the government because important economic reforms were being tried out, and were showing good results. Tragically the province (traditionally known as the Kingdom of Paradise) has often been held up as an example to other areas because of its good grain harvest and plentiful resources. But it is in danger of having its reputation ruined by the increasing threat and reality of floods.

Still, the Hwang Ho is the more serious menace to the Chinese people. So disastrous were the floods of September and October 1887 that 300 villages were swept away, 50,000 square miles inundated (the entire basin), 70-foot high levees overtopped and nearly one million killed. An additional two million were made homeless for years afterwards.

This is not to say that no anticipation, no preventative work, takes place during the flood season. The Chinese have had their universal flood memories (Chinese history, as we learned, begins with a universal flood) – spanning thousands of years – to learn from.

Jacquetta Hawkes, in her *History of Mankind*, talks of the Shang dynasty in the middle of the second millennium being confined to the rich loess plain on the north-west banks of the Hwang Ho, some 200 miles north of Peking. Protected by both the river and a long mountain chain, the area was highly suited to the development of a wealthy city state, with its timber reserves and level acres of grain pastures. On the banks of the river the Shang people built 'the Great City Shang'. By all accounts they were a highly vigilant people. As early as 2356 BC the river channels were dredged down to the Gulf of Chihli at Tientsin, and the first series of levees were built in 602 BC when it was clear that the flood threat was visibly worsening. It was in that year that the Hwang Ho first carved its way to the Yellow Sea, and confirmed ominous volatile characteristics that were to make the Yellow River the most infamous in the world.

As the death toll mounted with the passage of each century, more and more resources were devoted to diking, diverting and damming this ungovernable river. One hundred years ago the most flood-prone regions of the Yellow River Basin, some several miles from the delta, the river's banks were marked off into short sections, and each had its allotted river-watcher. So seriously did the Chinese take the flood menace that the river guards were in fact officers of the Imperial Army, highly experienced in disaster relief work. They had the authority to deploy men and materials to shore up crumbling dikes and organize further damming work.

Considerable sums of money were spent on vital repairs during the winter and spring when the flow was less fierce. Then, come the summer, great hordes of other ranks were employed to watch over the river's behaviour. Equipped with implements and diking materials, much of which was laboriously transported hundreds of miles, they were ready in an instant to strengthen any visible weakness in the bank. At the time of the 1887 flood, work was concentrated on Tsinam, the capital city of the province of Shantung, where the waterway narrowed but had a high volume. But in spite of this vigilance unexpected breaches still occurred without exception every time the inexhaustible snows poured down from the highlands. Wide tracts of land were frequently overrun, harvests ruined and peasant farmers made homeless.

But there was one other dangerous feature of the Hwang Ho that bode ill for all Chinese and presented the Imperial Army with its most indefatigable enemy. Every now and then the river dramatically and unexpectedly changed its course. While soldiers were painstakingly patching up a fissure or breach in the dike wall, the Yellow River would peremptorily desert its river bed, leap over the bank, burst through the levees and pour down a densely occupied plain, meteing out unparalleled death and desolation to a captive populace. From then on, until the next deviation, it followed the new course to the sea. Often a change in route meant an exit at a new river mouth 270 miles distant from the previous one; it was as if the river Humber suddenly decided to exist where the Thames is now.

This is what the Hwang Ho did in October 1887, and had done some five or six times in the previous 2,000 years. It had done it earlier in 1852, forcing itself into a new path some 300 miles across Shantung to the north of its former route, when it used to empty into the Yellow Sea.

The events that took place in 1887 were only partly the result of wet and stormy weather conditions, plus melting glacial ice from the Tibetan mountains. The Yellow River possessed another unique characteristic, or at least a more pronounced one. For some 500 miles the bottom of the river bed continued to rise markedly above the surrounding countryside. This is because of the perpetual dumping of its yellow sediment (loess) as it slackens speed on reaching the Great Yellow Plain in the north. This would not normally cause much of a problem if the river did not carry so much of the silt. But it is twice the amount that flows through the Mississippi. And each year it is reckoned that the river dislodges over 1,200 million cubic yards of soil. An old Chinese legend decrees that restless and wicked dragons stirred the mud in rivers in order to cause great calamities. Even in the early years of this century foreign engineeers working in China often had their work delayed while the river dragons were propitiated.

In some parts of the country the bed of the Hwang Ho rises by some six feet a year. The result is that at low water the river runs about 15 feet above the general level of the plain. To make things worse the floodwaters cannot easily drain away. Some regions stay inundated for a whole year. At high water the level can reach 30 feet, and in flood way above that. Almost annually some 3,200 square miles of China are inundated – to various degrees of severity – in this manner. Some Chinese scientists are talking about the danger of the Yangtse Kiang developing into another Yellow River, although at present the level of silting is less than one-seventh of the latter.

As a result the constant and repeated raising of the levees represents a never-ending struggle against nature. Many Europeans in China during the nineteenth century testified to this gruelling fact when they visited the old channel left dry in 1852 after the change of course, and found themselves walking uphill as they entered it through a gap in the massive walls. As they gazed at the open plain below upon which nestled tiny peasant hamlets, even those of the most sanguine nature realised the enormity of the flooding threat. Now the dikes tower like a miniature Great Wall some 24 feet above the surrounding plain.

This extraordinary set of geographical features was the backdrop against which the drama of 1887 was played out. In the latter half of September of that fateful year the river crashed against a bend a little below Chengchow, about 40 miles west of Kaifeng in the populous Honan Province. The embankment, already sodden with ten days' continuous rain and subject to howling winds, abruptly gave way under the onslaught. One hundred yards of embankment was torn away, together with a strong eastern wall situated just behind. The river continued to follow its channel until the breach had widened it to 1,200 yards.

But the principal exacerbating feature was the unusual coexistence of other smaller rivers in a vulnerable plain where the Yellow River approached its widest point. This was probably the worst site for such a breach to occur since, unusually, a smaller river, the Lu-chia, was at that point running parallel with the Yellow River. And the valley of the Lu-chia was much steeper at that point that the Yellow Plain. So when the flooding Hwang Ho took over the smaller river it had both the destructive volume and the necessary elevation.

The waters hence rumbled towards the walled city of Chungmou with astonishing force. In the surrounding district some 100 villages were entirely engulfed. Further on the paddy fields belonging to the inhabitants of 300 other villages were swamped. Chungmou itself survived the attack for some four hours, then finally succumbed.

The floodwaters then turned south, following the route of the Lu-chia, and a great tide of sepia-coloured water 20 feet deep spread out for 30 miles, added to by the small tributaries of the region. Chusien Chen, a sizeable trading centre of China, was fortunately on higher ground, and escaped with the loss of a few suburbs. Then a geographic multiplier went to work – so common in flood scenarios – where the damage already wreaked is added to by other forces.

Some 70 miles south of Kaifeng the Lu-chia and Yellow River together were joined by a larger river, already overspilling, from the west. At this point the population was living in the most low-lying and fertile flood plain and hence – another vital flood coefficient – the areas were densely occupied. In an area that could not have been more than 30 miles square as many of the 1,500 villages were submerged. In one town, named Cho-chia-kow, fifty streets were completely flattened, and the area beyond the town was submerged up to 30 feet, overtopping many peasant buildings.

It was to be several months before the final death toll could be guessed at. The course of the flood was about 300 miles, inundating 50,000 acres of crop lands to depths of 25 feet in some places. Eleven cities were drowned, along with about 2,000 villages, and millions were made homeless. There was a huge stream of refugees. A correspondent for the *London Times* wrote that the death toll 'cannot well be less than one million, and probably is not so high as two'. Although the general consensus of opinion puts the total number of fatalities at 900,000, there are many authorities who put the figure at anywhere from two to six million by including flood-related deaths arising from disease, fatal injuries and starvation.

As if the Hwang Ho and the Tientse Kiang in their natural states were not fearful enough, Man has added to the hazards. The high death toll in the 1938 flood was entirely due to the acts of Chinese soldiers who blew up dikes in an unbelievably callous attempt to halt the advance of Japanese armies. That was one occasion when the Yellow River was compelled to shift its course into new terrain. The dikes were dynamited on the orders of Generalissimo Chiang Kai-Shek, leader of the Nationalist Chinese government, to try and prevent the imminent take-over of the city of Chengchow.

There could have been no greater illustration of the maxim that life is cheap in the Third World, since the exegencies of war proved more important than the lives of the indigenous population. The floods destroyed eleven cities and 4,000 villages, and ruined the valuable crops of three provinces. It took the lives of nearly one million individuals, and left up to 11 million more homeless and without food. And all this for a notably limited military gain. The peasants still alive in the area remember the event bitterly.

Furthermore, the floods proved an annually recurring nightmare until 1947 when further dynamiting by the Chinese forced the river back to somewhere near its original path. For example, further catastrophic floods occurred in August 1939 when the Japanese were in occupation in many areas. Numerous rivers in northern China were affected when yet more dikes were blown in the Yellow River. Tientsin was threatened with the worst inundation of its history when the Hai River rose alarmingly. At various consulates in the city the water rose to ten feet. The British and Russian embassies were flooded out and plunged into darkness when electric power plants were damaged by water. In all, about half a million may have died as a result of the 1939 floods, and several million more may have perished as a result of the famine following the disaster.

The years 1938 and 1939 were horrendous ones for the Chinese. They were victims of probably the highest mass death toll arising from floods – anything up to five million – anywhere in the world. Unfortunately a great many of these deaths were the product of war rather than natural disaster, and heavy censorship and the absence of western correspondents have, even today, prevented the full details of these horrific and wanton events from being disclosed.

Why Men Live on the Floodplain

The overflowing river can cause much devastation to the human habitat because of the propensity to occupy the delta or floodplain. As Ronald Hewitt says in his *From Earthquake, Fire and Flood*, disasters were not disasters until Mankind arrived on the scene; they were simply 'nature's great dramas'. If anything, the sociological dimension has become more important as time has passed. The impact of weather and climate on people has increased because population numbers have risen. More buildings are erected, more land is farmed, and more food is eaten; so each traumatic hurricane, drought or flood has a commensurately greater impact.

Nowadays an estimated half a billion people live on floodplains, with untold

The emperor Franz Joseph of Austro-Hungary visiting the flooded town of Szeged shortly after the river Tisza burst its banks.

millions more living along vulnerable coastal areas subject to sea surges. Perhaps as many as 200 million Chinese live in the Hwang Ho and Yangtse River valleys. And the major reason why people remain in such vulnerable situations is, of course, economic necessity. Something like one third of the world's population is fed from produce grown in these fertile regions, so it is unlikely that patterns of risk will be measurably reduced in the near future.

Indeed, as famine spreads across those other regions plagued by drought the river valleys of the world seem more and more to be highly desirable places in which to live. The soil is often deep and easy to cultivate, containing deposits of rich alluvium, and can provide the basis of a thriving agricultural industry. Through one of nature's ironies, the vast slimy sea of liquid mud surrounding the banks of the Yellow River after a flood is an ideal fertilizer.

Then there is the ready availability of fresh water for irrigation, and the benefits to be had of river transport to facilitate distribution of merchandise, etc. Robert Raikes, in his book *Water, Weather and Prehistory*, has drawn attention to the discovery of abandoned river terraces precisely because of the return of drought conditions. In the developed world, too, there are aesthetic attractions to living in riverine locations (and beachside locations – in Melbourne and California many residences are built on stilts only yards from the seafront), which invariably carry a prestige price value. In the US it is reckoned at least 12 per cent of the population live in floodzone areas.

Even so, the advantages of settling along riverine sites was not so compelling for the ancient civilizations – the Romans, Sumerians and Chinese for instance – that they took a chance on survival. They clearly had a finely developed environmental instinct, and consciously chose sites for villages that avoided unnecessary risks of inundation. The idea was to build a settlement not *in* the river plain, but adjacent to it on a bluff or terrace. Paris, Washington and Bonn all lie on a series of high river terraces, originally well beyond the highest bank-high flows. The early Roman settlement of Londinium, for example, built on a series of hills (the present day City of London), was virtually floodproof. The Romans could hardly be blamed for not anticipating the massive growth of Greater London in the ensuing centuries, and the increasing settlement along the more flood-prone banks of the Thames to the low-lying east of the City.

This is why European floods have, throughout history, increasingly become social problems as the population has grown. For this reason the authentic history of floods in America is generally less than 200 years, after sizeable populations had become settled, compared with the centuries in Europe and Asia. Climatologists have records of French, German and Spanish rivers overflowing regularly since the twelfth century, but it was only from the fourteenth century onwards that the death toll from drowning became worthy of comment in ancient annals.

The medieval floods of the river Turia in Valencia, in the eastern province of Spain, were notorious. In February 1328 bridges were destroyed, and in August 1358, after a prolonged drought, 1,000 homes were undermined and collapsed. Another overflow in 1424 destroyed part of Valencia's town walls and over 600 houses, and played havoc with the agrarian economy. Later in the middle ages,

Floods in the Thames Valley in early Victorian times, when several monsoon-like summers frequently caused the Thames to overflow.

rainfall increased and shifting tectonics and other geological disturbances enabled rivers to soon achieve their present depths. There were disastrous Seine overflows in 1649 and 1658, and widespread disruption was caused by the Danube in 1501 and 1787.

The flood plain dwellers in the industrialized world naturally suffer less from flood risks than those similarly placed in the Third World. Not only do those in the latter suffer far more privations than those in the West – often grinding poverty and the periodic threat of famine – but they are now often the victims of natural meteorological excesses. We have seen that in 1970 nearly a million died in the Bangladesh cyclone, and at least 9,000 Hondurans were killed by Hurricane Fifi in 1974. In 1973 alone, 25 major disasters killed 110,000 people and disrupted the lives of 215 million more.

Unfortunately the inhabitants of Third World countries make things worse for themselves. Apart from the question of deforestation and soil erosion which exacerbates the flooding problem (see Chapter 10), the flimsy homes of the poor are highly unsuitable for countries subject to the violent wrath of nature – in particular the cyclones and earthquakes that frequently visit wide areas of Asia and Latin America. Furthermore the governments of poorer countries have fewer funds available for planning and preventative measures.

Since the War a great deal of flood prevention work – not to mention the more or less continual dike raising programmes – have been carried out in China, especially in Honan. Flow velocities have been slackened by special piers, baffles and dams. But the build-up of sediment is a constant threat to virtually any remedial measure. The Sanmen Gorge dam opened in 1960 could originally control 92 per cent of the water, but this capacity is already down to 50 per cent.

The Chinese government in the meantime continues to work on ambitious

Probably the worst French floods this century took place in 1910 when the Seine overflowed catastrophically. These pictures show the aftermath in Paris.

schemes to develop rivers with an aim of reducing the menace of regular flooding, including two large hydro-electric projects, one near Lanchow and one in the Tungkwan Gorge area. Tree planting has started, but it is a huge task. The Yellow River is an intractable opponent, and it is difficult to believe it will not get its own way with the land as it has done in the past.

In the West, on the other hand, the policy is to discourage people from settling on low-lying river deltas. And forecasting and other strategic responses to the flood hazard are becoming more efficient. But still the West is not immune from a flood menace that grows annually; not least, ironically, because of the sense of security such improved techniques instill in the minds of homesteaders. According to Maurice Arnold, a director of the US Bureau of Outdoor Recreation, earlier populations at risk were encouraged to return to the floodplain. This turned out to be a dangerous practice. Those regions which relied on structural means to combat floods had larger than average flood-risk populations, so the damage potential was higher. Local residents continued to drown in spite of all the warnings and all the precautions.

Left to themselves, people display profound ignorance of the most important geographical characteristics of their chosen place of abode. They have short memories, too. In the Hunter Valley, in the coastal region of Australia's New South Wales, very few new residents were able to respond accurately to a questionnaire concerning the likelihood of floods, despite the lengthy record of serious inundations in the area. Some respondents favourably cited statistical averages which, they believed, proved that the problem was insignificant. Instead, they should have searched for evidence for the *extremes* of climatic violence, as drawn to their attention frequently enough in the media.

John Whittow, in his book entitled *Disasters*, gives an example of this lack of awareness in a British study of 1971 carried out in Shrewsbury, Shropshire. More than half the residents living in a notorious floodplain of the river Severn did not expect a flood in the future. Most felt that a flood was the least of their worries, since such a hazard had to compete with the more familiar disamenities of, for example, traffic noise and vandalism. And if a flood came, most reasoned they would more than likely just move upstairs and sit it out. American studies at Rock Island, Illinois, confirmed the close attachment of residents to their riverside locations, and said they would be uncertain of their own reactions if faced with a flood hazard.

In the underdeveloped world the prevalent mood was apathy rather than complacence. Whittow cites the depressing case of the villages on the Ganges flood plain in India. They were quite aware of the flood hazard, but simply preferred to bear the loss to housing, crops and livestock than move to another area. They were fatalistic, like so many in the East, while at the same time placing the responsibility for flood management schemes onto the state.

Since the last war the flooding threat is being taken increasingly seriously in Britain. Like Continental Europe, Britain suffers from the hazard of dense urban areas that have grown up around meandering river basins. Most of these overflow with varying degrees of severity and with disturbing regularity. The twentieth century has brought an increase in artifacts of all sorts. Along with

Two rare prints of an unnamed Hungarian town before and after a serious river inundation. Probably late eighteenth century.

more dams and sluices have come telegraph poles, new gas and electricity stations and garages – all highly vulnerable to destruction by floodwaters. As communications spread – rail, automobiles, telegraph wires – so disruption was accordingly greater when the floods came.

Then there is the problem of Britain's capital. London is a major European city with a serious geographical defect rarely found in other similar conurbations – it has a waterway running through it that is more of a tidal estuary than a river. Some 30 years after the 1953 floods – after much traumatic haggling, designing and digging – the massive and unique rising-sector Thames Barrier was officially opened, and is now a visible reminder of the London authorities' attempt to keep the surge tide out of the city once and for all. A huge Doomsday Book was also hastily compiled by the National Water Council to pinpoint the most probable target areas for major flooding in England and Wales, and the likely chaotic consequences.

One of the chief hazards in Britain is the dangerous state of about 100 dams which need major repairs to the value of £30 million. Many of these are publicly owned, so because of economic stringency the Council feel aggrieved that little is being done to put things right. As a spokesman for the Council said: 'You can't prevent floods, but you can protect yourself from them to a certain degree.' The Institute of Hydrology now undertakes satellite surveillance hooked to automatic weather stations. Radar is also being used more. But there are problems involved in the use of modern methods, since the Meteorological Office itself still can't predict exactly how much rain will fall in specific areas.

A man abandons his submerged car after the Red Cedar River rose 16 feet above normal at East Lansing, Michigan, in 1975.

Late nineteenth-century floods in Sacramento, California, after the Sacramento River overflowed.

Chapter Eight

THE DROWNING OF AMERICA

America's Spring of 1983 was little short of a meteorological offensive; an onslaught of tornadoes, thunderstorms, torrential rain and blizzards. There were heavy snowfalls as far south as Georgia and Alabama, and flooding was so widespread that one journalist wrote of a need for a Salvation Navy.

But the malevolent spring weather was merely the tail-end of an extraordinary winter, marked by astonishing extremes of climate that saturated 350,000 acres of Californian farmland and threatened both coastlines with menacing 20 foot waves. Unusually, snowstorms reached as far south as Texas. And in parts of the western mountain regions more than 600 inches of snow fell, to bring yet further flooding when it melted in May. And yet the eastern half of the USA was basking in 70 degree temperatures over Christmas, while in Louisana flood-waters from 15 inches of rain kept thousands of people away from their homes.

But the tragedy of America in the eighties is that floods have now become a much greater threat to life and limb than the San Andreas earthquake fault. Many Americans doubt whether they can survive another winter like 1982/3 with its accumulating icepacks and heavy saturation. The trouble is that the country is all downhill from Canada. The terrain drops from nearly 5,000 to 500 feet in the space of 500 miles, coming down south-east to Oklahoma. Nowhere is the apprehension more acute than in Louisiana, where the Army Corps of Engineers is in mortal combat against gigantic odds in trying to implement a latter-day Mission Impossible: to control the Mississippi River.

During the winter the doomwatch scenario had already been painted vividly in Missouri, Arkansas and Illinois. When the Mississippi and Illinois rivers rose alarmingly, more than 36,000 people had to evacuate their homes as the floods spread a huge swathe of torpid water through hundreds of miles of flood plain. The Governor of Arkansas declared the state a disaster area after $450 million worth of damage had been done. The big rivers had been slow to reach their peak because of water thrust back and retained in the tributaries. It was the lull before the storm (or in this case the melt) because people in the three states and in low-lying adjoining areas had been piling sandbags on top of the levees. They were hoping to stem further floods, but at the same time were fearful that the

101

frigid weather would freeze the bags and render them less effective. Already the Illinois River was cresting 14 feet above the flood, and St Louis Harbour and much of the river up to Buffalo Island was closed to boat traffic for fear of wakes dislodging the sandbag defences.

The Army Corps has a massive task ahead of it. The Mississippi is the most important river in the country, and with the Missouri it forms the longest waterway complex in the world. It is 3,870 miles long, and is the country's dominating geographical feature. It rises in a lake in Minnesota and lumbers down the chest of America to the Gulf of Mexico, draining a vast area between the Rocky and Appalachian Mountains. It is fed by many troublesome, overspilling tributaries, such as the Ohio, Arkansas, the Red, the Illinois, the Yazoo and the Des Moines. At St Louis, where the Mississippi and Missouri meet, the river is one mile wide. The Indians, in awe, named it Father of Waters.

But as the weather turns increasingly neurotic this great river is threatening to run amok. Like the Yellow River it silts up easily, but also has to cope with the increased run-off rains and snowmelt not just from 41 states, but from Canada as well.

The consequences appal the imagination. It wants to take a shorter and steeper route to the sea. But if this happened it would bypass the ports of Baton Rouge and New Orleans and suddenly deprive them of the shipping facilities that handle 50 million tons of exports a year. While struggling to find a new course it would wash away roads and railways, rip out gas pipelines supplying 28 eastern states and flush out coastal communities, like the 50,000 inhabitants of Morgan City, out into the Gulf of Mexico.

A disturbing study from Louisiana State University recently drew attention to the present record flow of the Mississippi, and claimed that its volume was up nearly 250 per cent from that of 50 years ago. They pointed to earlier studies dating back to 1940 which showed that the lower 300 miles of the river has been trying to absorb yet another steeper river known as the Atchafalaya, which presently helps in draining some 30 per cent of the Mississippi to the Gulf. Every time the floodplain is swamped the flimsy network of levees is further eroded. In the massive floods of 1973 that submerged 11 million acres of land the river very nearly took over the Atchafalaya and thus change its course for the eighth time in 4,000 years.

The situation is complicated by the Old River Connection, which is an important waterway six miles long. It takes up a lot of the Mississippi's flow, and diverts it into the Atchafalaya via what is known as 'the old river diversion'. In 1956 it took nearly 24 per cent of the main river flow, and it has been estimated that it reached 40 per cent in 1975 and is now up to nearly 60 per cent. Herein lies the threat of a massive takeover by the Mississippi as the silting up and overflows become ever more frequent.

The Missouri, too, the Mississippi's smaller brother, has often caused much mischief to those living near its banks. Old maps and other sources show that the Missouri erodes and re-deposits 9,100 acres of valley bottom a year. It carries a high load of silt, even more than the Mississippi. This itself is silty enough, but not as much as the Colorado and the Rio Grande which can carry twice as much silt in high flood.

Yet the worst danger of siltation still lies with the Mississippi, known as 'The Big Muddy'. There is simply more mileage of silt to be carried over a much wider floodplain draining one and a quarter million square miles. And, like the Hwang Ho, the channel is remorselessly built up above the level of the surrounding country. The treacherous lower stretch of the Mississippi forms a gigantic funnel for more than 60 miles. The flood does not always become a towering wave-tide; it is sometimes a largely slow-motion phenomenon with the flood crest traversing only 30 miles in one day. Before the erection of levees the flood waters would roll majestically over 30,000 square miles. Now the Big Muddy has 2,200 miles of levees controlling its lower course, and in addition many towns and cities have their own encircling levees.

So there is a continual hazard to the plain dwellers' way of life whenever the levees are breached. The danger is greatly exacerbated when heavy and prolonged rain swamps the Upper Mississippi and Ohio valleys. And there have been appalling floods in the past as a result of the Big Muddy rising higher than its tributaries. Then a backflow is created and the local waters quickly burst their banks.

Should all the tributaries be in spate simultaneously the greatest of all US disasters would occur. As it was, the 1927 flood broke through the levees in 47 places, inundating 28,000 square miles of country and submerging the homes of three-quarters of a million people. An immeasurable expanse of liquid murk stretched from south-eastern Missouri down to the Atchafalaya Basin in Louisiana where over 900,000 live. The flood reached 56.4 feet in Cairo, Illinois. There was 10 feet of water in the business section of Montpelier, Vermont. The flood killed 313 people, uprooted and made refugees out of some 931,00 more, and caused nearly $300 million worth of damage at present prices. Overall, the lives of as many as 1½ million people were affected.

This serious event was unusual in that the waters were quite fast moving – up to 20 mph – and were whipped up by stormy weather, thus adding to the difficulties in evacuating residents in time. The streets of Arkansas City were dry at noon, but by 2 p.m. mules were drowning before they could be unhitched from their wagons. The entire area was under water from February to July, the effects of the aftermath being felt further down the river than those of other 'spring floods' which up to then had normally petered out by May.

They were not, however, in spite of conventional wisdom, the worst river floods in American hitory. But they lingered longer in the minds of Americans who had been sickened by the weather, especially after the New York State floods of the following November, when nearly nine inches of rain fell in one day in Vermont. A more conventional spring flood occurred in 1882 when a record 34,600 square miles were covered, compared with the 27,000 involved in 1927; but they were less traumatic in their effect.

America's Growing Flood Threat

Why is America's flood problem getting worse? The simple answer is that surplus water is no longer draining away via the earth's natural channels. Since

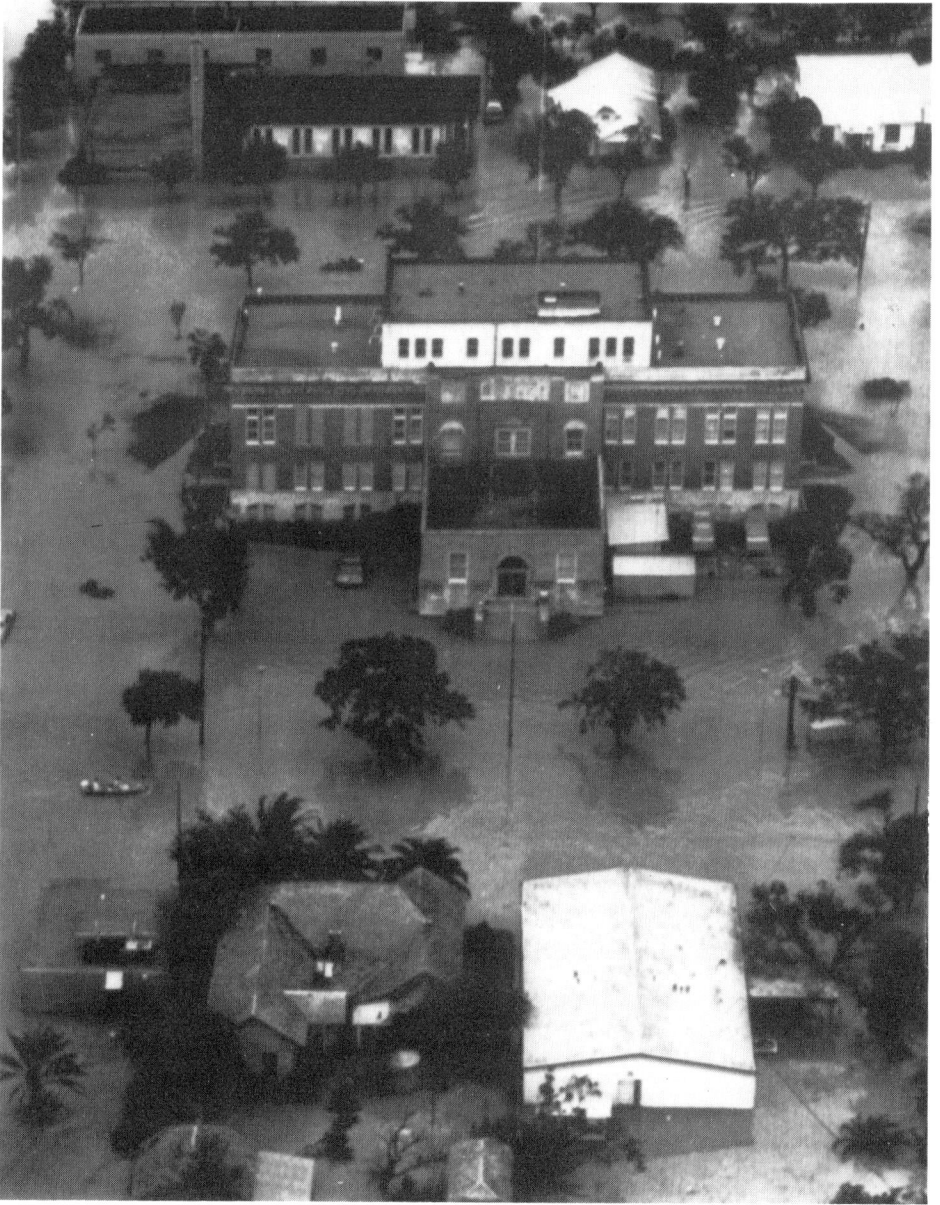

Fierce storm waters caused this flood in Sinton, Texas, in 1971.

the turn of the century enormous ecological and geographical changes have been wrought to the skin of the land. Only the Mississippi seems bigger than Man, and wants to change its course in spite of Man's desperate attempts to prevent it. In other smaller rivers, however, changes of direction are forced upon them by low bridge piers, locks, pipelines, sewer outlets, marinas and other obstructions. The rivers become effectively dammed, yet still seek urgent outlets.

Floodplains, too, are occupied by the infrastructure of civilizations – homes, public utilities, factories and of course highways (with their flat, hard surfaces, ideal water transporters). Even in the 1950s Holt and Langbein, in their seminal study of American floods, could write: 'These conditions are evidence of a growing and prosperous nation, but at the same time a nation not fully aware that there is no absolute assurance that future floods will not exceed the capabilities of our most comprehensive control measures, that our annual flood losses will not continue to increase, and that people who occupy flood plains will not do so at considerable risk. . . .'

Americans, however, can hardly avoid being at risk in a vast subtropical continent where all the prime causal agents of floods are writ large: giant sediment-carrying rivers, massive coastal storms and torrential inland rainfall. Furthermore, the zones of the worst floods are in the most populous regions – confined largely to the central parts, the east and south-east. Hurricanes thrust northwards from the Caribbean. The Mississippi-Missouri Basin (inaccurately known as the Midwest) meanders vertically from northern Minnesota down to New Orleans. Then there is the cloudburst rainbelt that runs roughly from western Pennsylvania south-westward into Texas; all those living within the vicinity of this broad crescent-shaped swathe suffering far more from fierce rain than other regions.

Of all these factors, the hazards associated with the Big Muddy loom massively and menacingly. The Misissippi Basin is not only huge, it embraces other major rivers that can all be awash simultaneously. This happened in May 1903, for example, when the Kansas, Missouri and Des Moines rivers all overflowed after heavy rains over Kansas City that killed 200 and made 8,000 homeless.

In few other countries do so many small waterways seem to overflow at the same time. There are some 3,600 rivers, forks, creeks, washes and bayous in the US on which rainfall discharges are being collated at about 8,000 locations. Many Americans are entirely at the mercy of these capricious creeks which seem to react entirely idiosyncratically to changes in the weather.

According to James Cornell in his *International Disaster Book*, river flooding is the most widespread geophysical hazard in the US, accounting for more annual property damage than any other type of disaster. The population living on lands virtually borrowed from rivers or the sea is over twice that of the national population density average. River overspills seriously affect at least 7 per cent of the total US land area, and are of major consequence for some 10 million people, and indirectly interfere with the lives of another 25 million.

Not all the US rivers are dangerous, as many have long since left their original floodplain or are considerably below the surrounding terrain. The Delaware

Early twentieth-century floods in Kansas after the Kansas River overflowed.

The overflowing Molalla River caused this disastrous flood in an Oregon small town in 1964.

Evidence of the US Army Corps of Engineers' success in preventing the Mississippi from changing its course to the Gulf of Mexico. The two control structures seen here are sited at the Old River complex.

River southward from Port Jervis to Trenton is a good example of a low-cut river offering little threat to the towns and farms in the area on relatively high terraces.

On the other hand, towns in the north-east are set in upland topography so rugged that the human infrastructure – towns, roads, railroads – must necessarily follow the valleys. And it is down valleys that storm waters rush. Some US citizens have their collective awareness shaped by a sense of being permanently in a state of hydrological siege. Like, for instance, the one million plus inhabitants of New Orleans. Whereas Venice is in danger of sinking into the sea, New Orleans has always been about eight feet below it, thanks to the French who built the city in such soggy surroundings. The city causes $19 million worth of problems every year just to keep it dry and upright. For a century or more the risk of building higher than four storeys was considered too great to contemplate. Even now the tall blocks of downtown New Orleans are erected on top of enormously deep concrete piers.

The people of Willard, Utah, too, are aware that their town has been built at the foot of canyons, and have often seen their down-sloping streets become tarmacadammed waterways.

But the other factor that is making the American flood threat worse is evidence that the hazards of unfortunate topography are being compounded by a

The rampaging Kentucky River sweeps through the towns of Hazard, top, and Lothair, bottom. This flood, one of the worst in east Kentucky's history, made more than 25,000 people refugees.

disturbed climate. Some experts are wondering whether more rain is falling on the States, and whether it is playing a disproportionate role in creating floods. Some historically notable floods are simply deluges, as in July 1916, when exceptionally heavy rain fell on the southern Appalachians. The Ohio River is especially prone to flooding because of the heavy rainfall on the Allegheinies. And in other areas any extra rain falling into narrow valleys, normally insignificant by US standards, frequently brings havoc and chaos. The fact is that much of the soil – apart from that in the arid west – is fertile and highly permeable, and weeks of rain can leave the ground saturated but disaster free. Then comes the final straw, when a violent cloudburst over waterlogged ground brings massive and often unexpected inundation. But good and ill fortune plays a major role. In the Allegheny and Ohio River floods of 1937 nearly 900 were killed by the rampaging torrents. The Weather Bureau estimated that 156 trillion tons of rain fell, making major and minor rivers overspill in 12 states. Some 12,700 square miles of territory were awash, more than a thousand people died through drowning, sickness and injuries, and 75,000 homes were demolished.

109

The Ohio spared not one major city near its banks. In addition all of the local rivers in Ohio state overspilled – the Scioto, the Muskingum and the Miami. The American Red Cross were said to have treated nearly 700,000 people in 1,700 refugee camps. The last time Ohio suffered as badly was in March 1913, when a typical sudden spring thaw accompanied by torrential rains had the entire valley network awash and took the lives of nearly 700. The worst effects were felt at Dayton, Ohio. This unparalleled flood in fact led to the nation's first flood control plan, then costing a massive $30 million.

And yet the most damaging and costly flood this century took only 41 lives in Kansas state in 1951. The people in the area were lucky. The flooded zone would have been one-third larger had the storm centred itself further to the north-west and thus into the Kansas River. After an unprecedented downpour of eight inches, the river swelled and surged through dikes, tore down four bridges, and forced more than a quarter of the population to flee Lawrence and the state capital of Topeka. As it was, most of the rain fell into the Neosho River.

America's most damaging flood this century occurred in the summer of 1951, when two great rivers, the Kansas and the Missouri, leaked into each other after a fierce storm. This picture shows Jefferson City, after the Missouri reached a width of six miles.

But the good luck was vitiated by bad luck. Kansas City had the dubious distinction of suffering from both fire and floods at the same time. This happened when an oil tank was torn loose and the contents ignited by a high tension electricity cable. Fires burned fiercely for nearly a week, driving an estimated 200,000 from their homes. Kansas is the largest wheat producing state in the Union, and the floods destroyed 25 million bushels, one-sixth of the harvest at that time. The swollen Kansas River leaked into the Missouri, and at Jefferson City reached a width of nearly six miles.

And we must not forget the legacy of America's mighty storms, the worst in the western world. The country's position in the northern hemisphere ensures that masses of westerly moist air surge across the American mainland, bringing with them cold anticyclonic breezes from the north. Half a dozen times a year Americans can expect, from the summer on, tumultuous hurricanes that pummel the southern Atlantic seaboard, which almost without fail devastate the Gulf and Atlantic coast states, while giant waves and high tides rip into the mainland. And virtually any area lying eastwards of the west coast will experience gullybusters throughout the often long, hot summer, causing local floods of great intensity. The Connecticut River Valley often suffers severely from violent inrushing storms, as it did in August 1955. For two days Hurricane Diane, following hard on the heels of Connie, dumped 14 inches of rain on the already saturated land of New England. Some 190 people died when a host of other streams burst their banks, causing nearly $1.8 billion worth of damage.

The year 1954 has been much commented upon in the meteorological literature for the devastation caused by two catastrophic storms. First Hurricane Edna deposited five inches of rain in New York in the space of one day, breaking a 45 year record. It wrought havoc in eastern Massachusetts, Maine and Nova Scotia. Coinciding with extremely high spring tides, the driving winds caused the sea to burst through defences to submerge vast coastal areas. A large number of docks and piers were put out of action, and hundreds of small boats and beachside homes were damaged or destroyed. As thousands of trees were uprooted in the raging winds they brought down miles of telegraph cables which in turn demolished other buildings.

Then, some months later, hurricane winds reaching 115 mph ruined much of the human habitat in southern Haiti before screeching on their way to the US via the Bahamas to become Hurricane Hazel – one of the greatest US storms in living memory. The full force of the hurricane hit North Carolina on October 17th before passing through Canada across Lake Ontario, leaving a 2,000 mile trail of devastation behind it. For a tropical storm emanating from the Caribbean it surprised many meteorologists when it penetrated Canada. Ontario and western Quebec received the full brunt of it. The storm tore through Toronto, and in one street alone 25 died. Cars and buses were hurled into a foaming River Humber, into which a row of houses had already collapsed, killing 19 people. Throughout North America tens of thousands were rendered homeless by Hurricane Hazel, bringing the death toll to 120.

This dramatic picture shows floods at their violent worst. Several times a year Americans can expect tumultous hurricanes to rip in from the Atlantic Ocean to cause fierce winds and torrential rains. Here an air force helicopter swings a man to safety from a beseiged house in Scranton, Pennsylvania, August 1955.

Snowmelt runoff is a frequent cause of the Mississippi–Missouri Basin overflowing, when spring temperatures can climb rapidly after a cold winter. This picture shows California Junction, Iowa, inundated after an unexpected warm spell, followed by heavy rainstorms, in April 1952.

The Big Melt comes to America

Evidence of more extreme and erratic weather emerges from the growing frequency of snowmelt floods. In June 1983 a late spring snowfall followed by a 90 °F heatwave gave Utah its worst floods ever, displacing 2,000 people and exacting some $200 million in damage. The transition from spring to summer is frought with hazards in North America, especially Canada. The long drawn-out British spring where the arrival of June often sees temperatures only five points up on that prevailing two months earlier is unheard of. Great sheets of ice, deep snow banks, frozen lakes, can all melt catastrophically within a few days, leaving nature to dispose of the excess water as best it can. The problem is that nature's best usually means humanity's worst. Severe flooding occurs when this phenomenon is more marked than usual.

The record floods of April 1928 in New York State were the result of a rapid thaw around the snowlocked Adirondack Mountains when temperatures rose up into the 70s after 30 days of weather at 8 degrees below freezing.

A similar flood took place in 1948 when the Fraser River overflowed to become the worst disaster in British Columbian history, indeed in Canada's history. The heavy snows of the previous winter were held in place by a prolonged cold spring. Then, at the beginning of May, the cold weather suddenly broke with disastrous consequences as the temperature continued to climb and remained high for several weeks. Vancouver was cut off in the

113

disaster. Some 55,000 acres of agricultural land was inundated and more than 2,000 left homeless. For more than a month the Fraser River Valley was in a state of liquid siege. Recovery was slow and costly as fields struggled to harden under several feet of mud and debris. Unfortunately the snowmelt flood is compounded by an important multiplier: lack of ground porosity. The frozen ground fails to absorb the meltwaters, thus facilitating a greater amount of run-off.

Other northern regions have their snowmelt problems, particularly Alaska, Siberia and other parts of Russia. The Rivers Ob, Lena and Yenisey freeze and melt disastrously every year without fail. In Britain the gravest snowmelt flood of recent times occurred to the end of the crippling winter of 1946–7 after widespread drifting of snow up to 15 feet deep in some parts of North Wales. But in America the main sufferers of melting winter snows are again the hapless residents of the Mississippi-Missouri system. In 1952 and 1965 the run-off caused by an unexpected warm spell was amplified by heavy rainstorms in the Upper Basin.

Rain is naturally the main agency in melting snows, itself being the product of higher temperatures. Rains invariably speed up a thaw, since they can melt snow even faster than strong sunshine, because much of the heat is reflected back by the surface of the snow. As a result of a crucial combination of rain and rising temperatures snow will on average thaw at the rate of about 250 millimetres per day, but only at about 65 millimetres per day under the influence of warm air alone.

Clearly a sudden thaw combined with rain often justifies a flood warning. But rain is a more important controlling agent in the snowmelt mix if the snow contains potentially about six inches of water. If the drifts are deeper the rate of thaw is the vital factor. Ice has one great advantage over snow from the human point of view – it generally melts more slowly. Glaciers for example are really part of the earth's heat balance mechanism and behave like some giant thermostat. They act as reservoirs for both liquid and solid water, retaining the winter's rain in the form of ice and releasing it gradually during the summer months.

As Europe gradually pulled out of its mini ice age late in the sixteenth and seventeenth centuries people living in the Alps suffered severely from melting mountain glaciers. The French historian Roy Ladurie, in his *Times of Feast, Times of Famine*, wrote that between 1600 and 1616, whole villages were destroyed. He mentions the Allalin glacier which dissolved into the Valley of Saas in 1663 and destroyed over 6,000 trees and countless peasant dwellings. Occasionally, however, providing the weather conditions are right, the glaciers too melt almost as rapidly as snow, as with the sudden release from the Tete Rousse glacier south of Mont Blanc in July 1892.

A continuing handicap for America is the notorious *ice jam*. Many rivers freeze solidly to great depths – measurable in feet rather than inches. Londoners have only experienced this during the mini ice age, when the River Thames froze over so solidly that 'frost fairs' were regularly held on its surface, complete with burning braziers!

Needless to say the ice jam under a rapid thaw is a highly dangerous phenomenon. The ice still gives the appearance of great solidity, but in fact has become deeply flawed with an invisible honeycombe arrangement of fissured ice crystals. Then, as the thaw continues, the ice breaks away from the banks, and the entire river begins to move alarmingly. Soon the water is racing clear from bank to bank, strewn with careening massive chunks of ice. It is when one of these ice slabs becomes lodged at a strategic river narrow or bridge opening that the hope is that the warming will continue. If it doesn't, and a freeze returns, the ice accumulates into a solid wall to await the next thaw; but in the meantime it causes greater than usual havoc downstream. Sometimes an entire section of the river will break up first, and small ice cakes will head downstream to pile up against massive re-freezing ice jams. The Missouri floods of 1952 were the product of this combination of events.

The Cost of American Floods

America, as rich as it is, would certainly have been a wealthier country without its floods. There are plenty of informed estimates to confirm this fact. Flood experts in the US Weather Bureau in 1964 estimated that flood losses amounted to more than $6,500 million during the period from 1925 to 1961, at 1964 prices. During this 36 year period financial losses in the Missouri Basin alone reached $1,797 million. D.C. Lacy and H. Kunreuther, writing in their 1969 book, *Economics of Natural Disasters*, were of the opinion that the damage from natural disasters excluding storm damage generally exceeded $600 million annually. Between 1950 and 1964, they believed, floods alone have caused almost $300 million worth of damage every year, in 1969 dollar values. If one included cyclonic damage the bill would have risen in those 13 years alone at constant prices to $550 million.

James Cornell, writing in 1976, believes that American floods now cost 60 lives and $1 billion in property damage every single year, compared with an inflation adjusted loss of $1 million at the turn of the century. Cornell points out that the actual damage cost often fails to include indirect losses such as the depreciation of agricultural land, disrupted commercial links or the long-term effects on the floodplain ecology.

According to Frank Lane, in his 1966 book *The Elements Rage*, the unenviable record for river flood costs goes to the Kansas-Missouri floods of June–July 1951. These floods, he said, were 'the only billion dollar floods in US history' (apparently $935,224,000 at less-inflated 1951 prices). He does, though, consider the price worth paying, since the death toll of only 41 was an eloquent testimony to the devotion and efficiency of the river forecasting and rescue services. Pennsylvania and New York State have yet to recoup the losses inflicted by Hurricane Agnes of 1974 totalling $4 billion.

As we mentioned earlier, America's flood history is only about 200 years old. Floods only became noticeable in the latter half of the eighteenth century and the first half of the nineteenth while the West was being won and southern California was being populated from former inhabitants in the Gulf of Mexico

Disastrous early Alabama River floods.

A family takes refuge.

region. Some of the nineteenth-century floods were truly disastrous. Holt and Langbein cite the floods in Iowa in 1851, the Central Valley of California in 1862 and the 1877 floods resulting from the overflowing Roanoke as being as great, if not greater, than those of the twentieth century.

From the late nineteenth century onwards floods became the subject of scientific study and measurement. By 1900 there was a network of raingauge stations across America, and the Weather Bureau began collating flood damage statistics. At the outbreak of the Second World War the Geological Survey had already commenced detailed studies of river flows, sedimentation and the cause and effect of each major flood.

There are now about 50 million acres of land in the US known to be below flood levels. Although this represents less than 3 per cent of the total, they are probably the most fertile, the most densely occupied, and the most economically active. Some 14 million are said to live in floodplains, including some 2.5 million in the Big Muddy's valley.

Those who live on coastal plains or beneath artificial dams are especially vulnerable. In no other country have there been so many reports of non-political refugees seeking succour and aid with such regularity (there are probably more flood refugees in China, but precise information is scarce and generally supplied by Western sources). For example, about half a million people had to evacuate their homes during the July 1951 floods on the Kansas and lower Missouri rivers. Three times that number were directly affected in January 1937 when the River Ohio rose 29 feet above the expected high after widespread and persistent rain.

117

The future, in the view of some experts, looks bleak. Raphael Kazmann, a civil engineering professor at Louisiana State University, wrote in the university's 1980 report that in about 30 or 40 years time at the outset, or even 'within a year or two', there could be an 'awesome' flood. Indeed, some scientists believe that whether or not the Corps of Engineers manage to complete their work on controlling the Mississippi by the 1985 deadline the battle has already been lost. The flow of 800,000 million gallons a day or more will prove to be an irresistible force. The future of New Orleans, as with the future of the world, will probably be determined by global weather systems which show disturbing signs of becoming more erratic.

Chapter Nine

THE FLOOD ENGINEERS

The Johnstown Dam disaster of May 1889 was, after the Galveston floods of 1900, the worst flood to befall America. A horrific combination of fire and surging water caused much terror and suffering to the inhabitants of a small town who happened to be living in the wrong place at the wrong time.

Sadly Johnstown, Pennsylvania, was no stranger to floods. People had often been forced upstairs in late spring when the rivers were filled with snowmelt. But in the spring of 1889 the snows of early April were followed by abnormally heavy continuous rains for nearly a month. Millions of tons of water were deposited over the South Fork Dam's watershed.

This dam, on the Conemaugh River, held back a large reservoir called Conemaugh Lake. But it was a highly dangerous edifice. It was then the world's largest earth-dam – 930 feet long, 80 feet high and 270 feet wide. In fact it was *too* large, and structurally flawed. Indeed, suspicions about the dam had been voiced long before it finally burst as a result of negligence and inaction.

The dam had been allowed to fall into disrepair some years after it was completed by the State of Pennsylvania in 1852. There had already been a break in the dam in 1862. And in the early eighties a brief attempt was made by some industrialists to strengthen it with tons of rock and mud, a fairly conventional practice in those days. Unfortunately Conemaugh Lake had been stocked with fish, and wire-mesh grates had been installed over the drainage channels which later became clogged up to cause overspills. To make things worse, the top of the earth spillway was levelled to enable a road to be built across the lake. This succeeded in lowering the spillway to only seven feet above the dam – representing an extremely low margin of safety.

This, then, was the poorly engineered and maintained structure that had to suffer the strain of month-long rains followed by about a further 16 inches that discharged itself from lowering skies on the evening of May 30th and which continued until noon the following day. By then the streets of Johnstown were already awash up to six feet in places.

The water in the dam rose alarmingly, failed to escape from the blocked overflow grids, and tore a gaping breach in the centre of the dam. Some

Artist's impression of the drama and aftermath of the Johnstown dam disaster, probably based on eyewitness accounts. Note the torrential rain through which the corpse collectors had to work.

120

observers said the burst was so explosive a blast of air threw down nearby young saplings. Four and a half billion gallons of water weighing some 20 million tons took nearly an hour to surge out of a gap nearly 300 feet wide. The debris collected on the way down the valley jammed the tide against walls, and for a moment slowed it. Then it surged ahead again, gathering momentum, demolishing stone and iron bridges, knocking out locomotives halted on higher ground waiting to enter flooded Johnstown, and tore down trees and houses.

The village of Mineral Point was completely destroyed, along with the Gautier steel mills and railway yards at Woodvale. The waters then surged through the township of Conemaugh and Franklin before the inhabitants of Johnstown became aware of a frightening low rumble. They were dead on target.

The tide ploughed through streets carrying everything before it until stopped by the town's centrepiece – the huge railway bridge with its 50 stone piers. This acted as a buttress for the flood's debris which towered around it over 70 feet high. The assorted accumulation of spoils from this disaster was an incredible sight: steam engines, carriages, trees, steel track, telegraph poles and carcasses of horses. Many hapless residents had already taken refuge behind the bridge and others were washed there against their will. Tragically an overturned freight car caught fire when fuel oil leaked, and many of the wooden structures and furniture became an inferno. The fire blazed with fury for three days, and about 100 people were burned to death.

The highest estimate for the Johnstown death toll was 2,280 – probably the worst dam disaster in history. It took over 7,000 helpers to clear up the mess, and it was several years before the town could return to normality.

Man-Made Aqualanches

One obvious fact that emerged from the South Fork Dam disaster – a sad truth at the root of most dam tragedies – was that the edifice was situated above a populated area, and the released water *fell* on the people below. True, the dam was 16 miles distant from the main built-up area, which sounds a lot. But it was also 400 feet above it. And no matter how porous the surrounding soil, such a deadweight of 20 million tons cannot possibly be absorbed. On the contrary it gouges out the land, and this detritus adds to the battering force. And whatever water is lost on a 16 mile downhill trip is vastly compensated for by the assorted flotsam collected on the journey. Over such a distance the torrent would turn into a landslide of mud, rock and timber.

Another example of a dam burst from a steep elevation was the Dolgarrog disaster of 1925 in which a tiny Welsh village, located on the floor of the Conway valley, was destroyed. A mountain lake in Snowdonia was enlarged in 1910 to provide hydro power to a new generating station lower down the valley. Following prolonged rainfall the insecure foundations were undermined causing the dam to disintegrate. The powerful floodwaters tore out scores of boulders, some weighing up to 500 tons, and it was these which largely brought about the high death toll.

Let us look for a moment at the worst dam flood in British history: the Dale

Dyke Dam disaster of 1864. The prominent feature of the Dale Dyke Dam was its natural configuration, as it was little more than a long valley filled with millions of gallons of rainwater and stopped up at one end with a wide barrier of earth, clay, timber and rocks. This was not the only similarity the dam had with the South Fork Dam, as it was of similar size and depth, and had just had to cope with a harsh season of above average snow and rain.

A warning was passed on about a possible disaster after the reservoir embankment sprang a serious leak. In fact a commendable attempt was made by senior engineers to save the dam, including the use of explosives to create a diversion and release the growing pressure. The attempt failed, and a portion of the embankment 110 yards long gave way abruptly, to allow the water to burst thunderously down the hillside. For several hundred yards the torrent tore deep gorges in the valley, parts of which were solid rock.

After demolishing the Yorkshire hamlet of Lower Bradfield the aqualanche fed on itself. Like a victorious army, what it lost in impetus it gained in destructive reinforcements when new territory was conquered. Villages at the bottom of the steep valley were scattered on ribbon-like slopes of low land, a ready target. The waters raged through any available crevasse, blowing out doors and windows. Many houses simply collapsed on their occupants, most when loose beams of timber battered enormous breaches in walls, weakening the rest of the structure. Nearly 250 people died in this tragedy, largely because of the density of overcrowded houses in close proximity to each other.

Why Dams Collapse

The building of the reservoir is one of the civil engineer's biggest responsibilities and his greatest service to his fellow man. A dam failure, resulting in a curtailment of essential water supplies and possible loss of life, gives rise to lengthy and tortuous public enquiries and even criminal trials. Futhermore, in no other branch of civil engineering is geology such an important consideration. Of the dam type that siphons off a river or lake the engineer cannot avoid interfering with geological processes. A pool is created and boxed in so that upstream sediment is deposited in the artificial lake behind the dam.

The water flow below and above the dam is extremely slow moving, and is unable to shift the sediment fast enough. The river too frequently becomes choked and overspills during periods of heavy rainfall. Within the reservoir itself the build-up of silt can become serious. Without constant dredging the construction becomes a potential water bomb, threatening to undermine and explode even the most soundly built embankments.

The building of dams has a long lineage. There is evidence of reservoirs and conduits in Babylon as early as 4,000 BC. Parts of the water supply system, mentioned in the Bible, are still being used to supply the Mosque of Omar. Notable historic waterworks in Jerusalem, installed when Solomon first conquered the city in about 900 BC, were skilfully made. Solomon's dams were extended by King Hiskia in 700 BC, requiring the excavation of a tunnel almost a mile long. When the Romans came they augmented the system to the extent of linking nearly every house in Priene, Asia Minor, with earthenware pipes connected to mains fed from a large hillside spring. Other ancient dams include the one built by Joshua to help the Israelites cross the river Jordan, and the first masonry dam built by Menes, the first Egyptian dynastic king, some time before 4,000 BC.

And it was in Egypt that the earliest known dam disaster occurred: some time between 2950 and 2750 BC the Sadd-el-Kafara Dam failed during its first flood season. A bigger dam failure dates from about 1700 BC near Yemen, when a dam wall some two miles long and 600 feet wide collapsed in a flood in AD 300.

In the ninteenth century the dam disaster was much feared. It was a century in which reservoir building was forging ahead to meet the growing water requirements of an expanding industrial society, and in some cases to bring infertile and wasting lands under cultivation. But the technology involved in their erection was often faulty or badly executed largely because of the limitations of geological knowledge. In those days few dams were the entire product of high-grade civil engineering practices. The rock-fill dam was commonplace because it was thought uneconomic or unnecessary to build it of more substantial materials such as concrete. The sheer size of some reservoirs precluded this. Most of them were little more than man-made lakes.

The civil engineer, Robert F. Legget, devoted much attention to the phenomenon of dam collapses in his classic *Geology and Engineering*, published some two decades ago. He pointed to two main factors: inadequate spillway capacity (or, in other words, insufficiently large overflow outlets within the dam itself), and defective foundation-bed conditions. A major hazard arose from the highly permeable foundations upon which some dams were built.

In his *History of Dams* Norman Smith believes the collapse of the Dale Dyke Dam was 'initiated by the faulty positioning and laying of the outlet pipes'. The heavy bank was already water-sodden and began to subside and dislodge the pipes, and water washed away the puddled clay surrounding them. Soon the water had eaten away the interior of the dam. Then the central portion of the dam slipped and the overfull reservoir spilled over the poorly constructed crest. The dam, now eroded from above and below, inevitably collapsed.

Many other dams have collapsed because of the absence of grouting techniques which allowed water to undermine the foundations. Even in the twentieth century this was a common cause of failure. It has been estimated that there have been 100 disasters arising from water eating away at crucial base areas to weaken the rest of the structure. This was said to be the cause of the collapse of the massive concrete St Francis Dam in California, in March 1928, which drowned some 400 people.

It has been reckoned that there were 60 failures of other geologically undermined dams alone between 1869 and 1919. Classic examples included the failure of the Puentes Dam in 1908 and the Austin Dam, Texas, at the turn of the century. This particular edifice, when finished, looked impressive. It was over 1,000 feet long and 68 feet high, and made of masonry. But it rested on Cretaceous limestone and clays. The limestone strata was highly variable, some parts being softer than others and liable to subside without warning. In addition

A burst reservoir caused this flood in Pavia, northern Italy, in 1895.

the whole structure rested on a fault zone. Frequent bad weather and high rainfall had already caused overtopping to erode some of the limestone, and on 7th April, 1900 a huge central section simply collapsed.

Throughout history the failure to either understand or predict often volatile meteorological factors has been a crucial contributory factor in dam disasters. Legget says, for instance, that the size of the overflow outlets in dams are related to the *anticipated* run-off, and from the size of the catchment area above the edifice. Seldom, however, was a wild season of weather anticipated: there was always insufficient leeway. Spillovers did perhaps allow the run-off for a week or two of excessive rainwater, yet this was often dismally inadeqate. Today the engineer feels compelled to allow the maximum runoff possible to cope with months of inclement weather, with all the statistical extremes of rainfall and storm damage taken into account. And he would include all the relevant, and vital, geological considerations. He would have to know exactly the nature of the land surface, and subsoil, he was building on.

And yet dams still collapse largely through professional errors of judgement. On 14th December 1963 a dam fractured because it was sited over the San Andreas fault in Baldwin Hills, Los Angeles, and water began to seep underneath a faulty retaining wall. Similar factors caused the disintegration of

125

The Vaiont Dam, Italy, looming 873 feet above the river Piave, was the third highest concrete dam in the world. A landslip from the surrounding mountains caused a giant wave to overtop the dam in 1963, killing nearly 3,000.

126

the newly built Teton Dam on the Snake River, Idaho. Indeed it was breached even before it was officially opened, causing $500 million worth of damage and leaving 30,000 people homeless. But the failure in this case was exacerbated by the unorthodox earth-filled construction built on soils which lacked stability in a geological fault zone.

Then there are the mammoth hydro-electric projects that are so huge that they become earth-shapers. They are executed with enormous panache, with billions of tons of concrete used to block up entire valleys. In the north-eastern corner of Italy, where the Italian Alps merge with those of Switzerland and Yugoslavia, there are many rivers that can be harnessed to provide much needed energy. When the Vaiont Dam was opened in 1960 it was hailed as being the third highest concrete dam in the world, its wedge-shaped wall looming 873 feet above the Piave River below. When it burst on the night of 9th October 1963 the waters plunged down the slopes of Mount Toc and utterly destroyed the village of Longarone, killing nearly 3,000.

But in this case the dam did not fail or collapse. The subsequent court of enquiry concluded instead that the dam was badly sited. There had been concern over Mount Toc for some time, as seismic tremors had been recorded. Even when the great walls were being constructed a series of landslips and cracks had appeared in the mountain, and instruments recorded stresses and strains in the rock face. What caused this disaster was a landslip from the mountain which dumped millions of tons of rock into the reservoir, thus causing a giant wave to overtop the dam.

A later Italian dam disaster, occurring only 30 miles from the site of the Vaiont Dam, was also said to be due to a combination of geological ignorance and the negligence of maintenance workers. This time over 230 people, mainly holiday-makers, lost their lives as a typical picture-postcard scene of red-roofed chalets and flower-filled meadows was soon turned into a great muddy lunar landscape.

It happened near the Austrian border in Northern Italy at a hamlet called Stava in July 1985. A split-elevation dam, built conventionally of earth and rocks, housed water used to purify Fluorspar, a powdery mineral mined nearby and used in steelmaking. Even before the disaster the fears of local people were due to be voiced at a council meeting. Much criticism was focussed on the decision to build a second reservoir for washing the mineral ore. Its weight rested on the basin of the earlier dam. Much of the blame was apportioned between the provincial administration and the technicians employed by the firm operating the mine, Prealpina Mineraria S. p. a., who failed to regularly clear out the muddy deposits that continued to build up.

In the words of one expert the dam was a 'geological time bomb'. In particular, Floriano Villa, head of Italy's geologists' association, blamed the dam failure on the lack of safety checks. Michele Jamiolkowski, a geology professor at Turin University, said the dam collapse was the result of gross negligence. An alternative theory focussed the blame on natural phenomena, such as water from a subterranean source, which had upset some precarious geotectonic balance. Even this theory did not, in some minds, exonerate the technicians working on

127

Part of the desolate aftermath surrounding the Italian village of Longarone following the Vaiont Dam floods.

This view highlights the dangers of building massive dams in seismically active regions. Mount Toc is in the centre, to the right of the valley, and the devastated village of Longarone lies immediately beyond.

The split-elevation dam at Stava, Northern Italy, before it burst in July 1985, killing over 230 people, mainly holiday makers.

the reservoir only hours before the structure burst, since any subterranean flooding would have been obvious.

Indeed, criminal negligence was suspected. The public prosecutor's office in Trento issued indictments to many of Prealpina's employees, including company executives, municipal administrators and all of the district's mayors for the previous ten years.

Often dams are simply defeated by the raging elements in spite of everything. The sheer wear of coping with climatological excesses can weaken the most solid of structures built with the sturdiest foundations. The Rapid Creek 500 foot dam in South Dakota, erected in 1932, finally gave way after receiving 14 inches of rain in less than six hours. But it had already survived three earlier floods, and thorough checks for structural or geological faults were carefully made after each event. Its spillways had also been cleared of silt, so it was a major surprise when it failed in 1972, knocking down 80 blocks of buildings and drowning more than 200 people. In this case the local residents took the 'blame' when they failed to take action when warned of the impending high water and dam collapse.

Some social critics view as disturbing the postwar trend towards massive hydropower schemes, claiming that most of the big dams are built for the benefit of the city planners. They claim there is an urban-industrial bias which enables designers to turn a blind eye to the fate of the fertile valleys that are frequently flooded out to displace large rural populations. The Aswan High Dam, for example, caused 120,000 people to suffer from serious floods, and the Koussou Dam another 100,000.

We are still living with the legacy of some appalling dam collapses. The memory of the Dale Dyke disaster has, even today, not been forgotten by Britain's top civil engineers. Surprisingly, more than half Britain's dams are more than 80 years old, and most of these are earthfilled. One estimate puts the total of these edifices still standing at 85 per cent, and most are considered to be in a dangerous state of disrepair. But no one knows exactly where they are. All around Britain there are dams which have never been inspected. Under an old Reservoirs Act of 1930 (largely spurred into being by the Dolgarrog dam disaster of 1925) they were supposed to have been checked out every 10 years or so by a panel engineer. But because of bureaucratic re-organisation the nation's flood problem is devolved to 66 local councils instead of one body. As problems mount, and as government spending cuts take effect, the later Reservoirs Act of 1975 remains unenforced, so maintenance of dams has been gravely neglected. The Institue of Engineers wearily highlight the sociological problems of many drains and reservoirs and canals lying above villages that have expanded into large communities. There they remain – still awaiting their fate.

Building Dams and Controlling Disasters

Whereas the problem of the nineteenth century was one of horrific dam bursts, the dam hazard in the present century has assumed a different, and awesome, dimension. These great expensive structures are, of course, built to supply energy as well as water. But, as with the Hoover Dam, the pursuit of economic growth and cost efficiency brings about other unreckoned and sizeable ecological disbenefits. Nature is being tampered with on a grand scale. The River Platte alone is now dammed 42 times across Colorado, Wyoming and Nebraska. The Soviet Union, too, has displayed such a mania for hydro-electric dam building that the Volga now looks like a 2,300 mile series of connected reservoirs. The Caspian Sea, fed by the Volga, is in danger of drying up.

The US Corps of Engineers, struggling to prevent floods in Mississippi, think little about the ecology of Northern Maine that could be destroyed by the construction of the proposed hydro Dickey-Lincoln Dam. According to biologists Paul and Anne Ehrlich, of the University of Stanford, the tiny snail darter fish is already threatened by the Tellico Dam in Tennessee.

And the Dickey-Lincoln Dam would ruin the habitat of bald eagles, bobcats, otters and moose. Its construction would entail the flooding of almost 140 square miles of valuable timber. And yet the Erhlichs think the benefits negligible: 68 permanent jobs, and the substitution of just one half of one per cent of New England's oil consumption. All over the US, they wrote in *Extinction* (1982), dams not only flood out natural populations but also divert waters from their original courses. It is in moister areas that diverse biological activity occurs, so inundating riverine haunts also extinguishes the lives of many species. More than 3,000 miles of irrigation canals built by the Bureau of Reclamation have succeeded in drowning numerous types of birds, coyotes and deer.

In the meantime thousands of residents of America's south-west and many Mexicans were forcibly reminded in July 1983 that it is not just wildlife that is threatened with drowning. Federal officials unleashed what they called a 'controlled disaster' on the West because all of the dams along the 1,450 mile long Colorado River were so swollen that to have failed to do so would have meant a far worse 'uncontrolled' disaster. Although Nevada's governor, Richard Bryan, was reported as saying that someone had committed a 'monumental blunder', the giant 725 foot tall spillways of the mighty Hoover Dam were jacked open for the first time to release nearly 2,000 tons of water a second. Failure to have done this, it was believed, would have meant that water from Lake Mead would either have overtopped the dam or would even have burst with a thunderous roar.

As it was, the Colorado – lifeline for seven states – went on the rampage. Three times the annual average of snowmelt from the Rockies flowed through the new opened flood gates of numerous dams in four states. Tens of thousands were obliged to become refugees while Mexican peasants and Indian tribes near the delta on the Gulf of California were washed out of their lands by this over-managed river. At least seven drowned, and the property bill was over $100 million.

The decision to open the dams, taken by the Bureau of Reclamation in Washington, was criticised by many expert hydrologists who said that the operation was undertaken far too late. Inhabitants of the flooded river towns could see for themselves how fast the water level was rising as the snowpacks rapidly diminished. Others made the surprising suggestion that political reasons played a major part. As in other parts of the world, reservoir water was jealously guarded. Often resentments were engendered whenever, by contractual agreement, water was funnelled to other regions that were considered to be rather more profligate in their consumption patterns. Welsh nationalists, for example, quibbled about local water taxes because a great deal of Welsh water was siphoned off for the West Midlands area.

Similarly, western states feared California's thirsty and expanding agri-business farms that had led to the building of an assortment of aqueducts, reservoirs and dams to divert the Colorado's water thousand of miles from its natural course. This had led to some counties endeavouring to protect their supplies, but at the risk of overtopping.

In addition it was believed that the bureau was under pressure from the Washington administration to economize and raise revenue sources. So flood control advocates within the bureau – who would necessarily want early opening of the dams to reduce levels – were in the minority. After the event the dam floods were generally considered to have caused irreparable harm to the Grand Canyon ecology. Health authorities had to warn of mosquito-borne diseases which had already affected the Mexican poor. President Reagan declared a six-county federal disaster area.

But the Hoover Dam fiasco is just one example of pressures placed on authorities to redistribute water by giving nature a helping hand. There have, in fact, been myriad professional rain-makers in America who have been only too successful. In his book *The Cooling*, Lowell Ponte describes how, in 1916, one Warren Hatfield undertook to end a drought at the request of the city of San Diego, California. With a strange Heath Robinson device he belched out smoky chemical fumes which caused 20 inches to fall downwind of the city, and washed away a dam in a flood that killed 17 people and did much damage.

Ponte also mentions the extensive use of silver iodide cloud seeding from aircraft to increase precipitation. He suggests that the violent Rapid City flood of June 1972 was the handiwork of a small cloud-seeding plane. Relatives of the victims of the flood sued the government in a class-action suit seeking $600 million and official acceptance of responsibility for the flood. The floods that hit North Vietnam in 1971 were also, implied Ponte, the result of American seeding operations.

Making Rivers Run Backwards

Yet the damming of American rivers, and the seeding of American clouds, might be considered an irrelevance in the light of what is happening in other parts of the world. Some of the mammoth late twentieth-century projects are so stupendous that they threaten to irreversibly alter the world's climate. Some authorities, of course, think this might not be such a bad idea. The Russians once wanted to purposively raise the air temperature in order to turn frozen Siberian tundra into arable land. A proposal was made to erect a dam across the Bering Straits to separate Russia from Alaska. The aim was to actually pump out the cold Arctic water, so that it would be replaced by warmer water from the Pacific. A Moscow meteorologist, P.W. Borisov, calculated that in only three years all the Arctic ice would have melted, and the basins of the Volga and Don would become subtropical.

For good or ill this scheme appears to have been dropped (one disadvantage would have been the dramatic cooling of the North Pacific which might have ruined the rice crop of Japan). But what of the latest Soviet scheme, now under way, to divert the great Siberian rivers that flow south, and so irrigate the increasingly parched and arid lands of central Asia such as Uzbekistan and Kazakhstan? But the risk is that the frozen seas of the Arctic will be reduced, ultimately causing droughts in the very regions of Asia to be irrigated.

This, at present, is just theory. Yet there can be no escaping the reality of such a huge landscape-changing exercise; probably the biggest ever attempted by man. Millions of acres of northern land will be drowned, including great tracts of game forest. Hundreds of historic monuments will be submerged, and tens of thousands of people will be driven from their homes. Towns and villages in an area larger than western Europe would disappear, some of them including the onion-domed churches dating back to the Middle Ages. There are fears that the historic part of Arkhangelsk would be closed off. It could also wreck flourishing fishing industries by denying salmon and other river species their freshwater spawning grounds.

Started in April 1983, the project is so vast it will take fifty years to unfold. But when complete a dozen rivers now flowing northwards into the Arctic Ocean will be flowing the other way, a seeming geographical impossibility. The Volga, as massive as it is, and all its tributaries, will have an even greater volume, and ultimately disgorge great cataracts of water into the Caspian Sea.

The first stage will be accomplished slowly. Some 25 dams, like giant canals, will be erected along the Onega, Pechora and northern Dvina rivers. The rivers will then pour into these dams, with the water level rising year by year until they are some 30 feet higher. Then the northern end of Lake Onega will be blocked off, causing the rest to pour out southwards, swallowing up five smaller lakes, before pushing through the expanded canal complex into the vast Rybinsk reservoir. From there it will be coaxed into the Volga.

Up to then it is estimated the scheme will cost £18,000 million. But then the second stage will attempt to reverse the flow of the massive Yenisi and Ob rivers in Siberia. This is when the demands on the engineers concerned will become colossal. Enormous dikes will have to be built across the mouth of Onega Bay to block off the northern outlets of various streams from the waters of the White Sea. Within a decade it is hoped the Onega Basin will become fresh. Soon after, the Divina and Pechora Bays will be tackled.

The original intention was to push the water south by building a canal some 1,500 miles long, perhaps by nuclear blasting. But that proposal drew vociferous protest from the West. Instead Soviet planners will now probably complete the second phase by re-routeing the water along old river beds revealed by satellite photographs.

The Soviets are convinced their grandiose river-reversing exercise will yield at least $40 billion. They predict a great improvement in grain production by as much as 60 million metric tons a year, up to 35 per cent of Russia's current crop. They also eagerly anticipate a halt to the rapidly depleting Caspian and Aral, the two major inland seas, because of the growing demands of irrigation.

However, it will be a long time before hydrological effects become noticeable. The Russian economist, Vladimir Perevedentseve, charged with predicting the full long-term effects, has asked to be taken off the project. In any event the planners will disregard climatic speculation. There are fears in the West that the increasing salting-up of the Arctic Ocean because of the future shortage of fresh water as the project gets under way will cause its freezing point to drop and the ice caps to melt. Other scientists fear an opposite scenario: as the flow of warmer water is reduced, the polar ice field may expand. It is possible, too, that future generations will suffer from the thick ice on the new reservoirs which will put back the spring thaw and shorten the growing season by two weeks. There could also be fierce winds and serious floods in the autumn.

In Conclusion: Why is Bangkok Sinking?

During the monsoon season of September 1983 it became clear for the first time – it had hitherto been suspected – that Bangkok was sinking. And Thailand's leading town planners were mostly held responsible.

Bangkok's flood problems have their genesis in the years following World War II, when economic growth lured hundreds of thousands of workers to the city. The 200 year old capital had begun life as a trading village set in marshy lowlands on the banks of the Chao Phraya River. A network of natural and man-made canals – *klongs* – provided transport and natural drainage. But as the city prospered, many citizens bought cars and took to the cramped roads, so the town's planners began to search for extra road space.

It was thought a brilliant idea to fill in the canals and cover them with tarmacadam. This turned out to be an unmitigated disaster. The city soon began to experience severe flooding every year. For years the klongs had acted as natural flood control mechanisms during the monsoon season – from June to October – when up to 45 inches of rain can fall. High tides or excessive storm waters simply backed along these deep and convenient waterways. Even in a bad flood, it was always easier to get out of the city when flat-bottomed boats plied the klongs. Now the city frequently comes to a standstill as the streets rapidly become waterlogged.

But why is Bangkok also subsiding? This time both the city's planners and its industrialists are to blame. The water supply has failed to keep pace with expansion, and the klongs provided a ready source of drinking water. So private industries, hotels and housing estates started to sink their own artesian wells deep into Bangkok's water table. When the quality of the water began to deteriorate, the wells were drilled deeper.

In just 15 years more than 11,000 of these wells have drained the underground reservoir dry, and the city is sinking at the alarming rate of about six inches a year. Parts of the capital have dropped by as much as three feet, and Prinya Nutalaya, of the Asian Institute of Technology, fears that as the land is only about three feet above sea level the entire city will be under water by 2001.

In the meantime residents have begun to notice pavements are subsiding. One shopping mall needed new steps to bridge a massive cavity at the base of a wall. Cracks in office walls appeared. Today Bangkok is fast living up to its name of 'the Venice of the East'. Floods are now much more likely than before if rainfall exceeds a inch an hour.

At last the authorities were goaded into remedial action. No longer could public apathy and the government's traditional *mai pen rai* (never mind) attitude be allowed to prevail. A master anti-flood plan was revealed by the Deputy Prime Minister, who said it would be completed within about eight years. The shortage of water would be overcome by piping it in from a river some 60 miles to the west. The private wells would be shut down, and a dike would then be built around the city, with a runoff canal feeding into the sea. Finally a desperate attempt would be made to stop the city from sinking by refilling the sagging water table.

Unfortunately work has not yet started on this imaginative project. There is one drawback: the final bill could run to several times Bangkok's annual budget of $207 million.

Chapter Ten

MAN THE FLOODMAKER: SLASHING, BURNING AND MELTING

The most serious environmental problem now facing the earth is *deforestation*. Its effects, if unchecked, are almost certain to bring about permanent ecological harm with a dramatic increase in the world's flooding problem. Quite possibly it will also bring about long-term climatic change. What is so alarming is the rate of tree felling around the globe that now is so rapid it seems highly unlikely that reafforestation will ever be able to make amends for the harm that mankind is doing to the skin of the earth.

Let us consider the facts. The world's entire forestry and jungle reserves now cover only a quarter of the land's surface, with the highest land ratio given over to South America, followed in order by Europe, North America, then Asia, and finally Africa. But up to half the world's original woodlands have vanished since 1950. It is estimated that a forest the size of Cuba is being destroyed each year. Brazil's forests have already been reduced by a quarter or more. Sizeable parts of South America, central Africa and the foothills of western Asia have now taken on the characteristically arid look of the Middle East. Even Britain has only 10 per cent of its original primeval woodland left. And in just the last ten years Thailand has lost a quarter of its forest; the Philippines one-seventh in the last five. At this rate much of the far east will have no lowland rain forest left at all by the year 2,000.

The size of the problem can only be understood from an historical perspective. Some 5,000 years ago the world would have been unrecognisably densely wooded, with most human activities seemingly conducted in large clearings set within interminable forestland. So familiar are we with the desert regions of the Levant that we find it difficult to believe the Bible when it mentions 'the forest in Arabia' (Isiah 21:13). The land of Canaan was fertile and full of 'vineyards and olive-yards, and fruit trees in abundance' (Nehemia 0:25). Indeed the famed Cedars of Lebanon remained until World War I when the last few acres were decimated by the Turks. Long before that the Pharaohs, the Phoenicians and the Babylonians had, over the centuries, plundered millions of acres of rich pines, junipers and oaks.

The Sumerians, in particular, were master tree-fellers. Their activities resulted in so much silt entering the Tigris and Euphrates and pouring down to the Persian Gulf that the two rivers have been pushed 130 miles further south. In Mesopotamia, the cradle of civilization, erosion soon took its toll when the dense forests on the Armenian hills to the north were chopped down.

Plato, speaking of Attica in Greece, said that the mountains were covered with trees, and roofs made from their timbers are still in existence. There were seemingly endless pasture fields, and the rainfall, instead of draining away, was

Tree-felling is an activity that has been undertaken for centuries in Europe and the New World. As a result the world's flooding problem has grown worse as nature's natural flood barriers are removed.

trapped and stored in impervious clays to be released later as springs. Reference to Roman writers clearly shows that certain European rivers such as the Danube and the Seine flowed much more sedately than they do now. And there can be little doubt about the reason: the later removal of forests that used to regulate run-off, and then the erosion of topsoil.

By about AD 900 central Europe north of the Alps consisted of at least 80 per cent forest, whereas by AD 1900 this had been reduced to 24 per cent. It was the areas of greatest population density that naturally made the most claims on the environment. Before the Industrial Revolution the chief building material for an advancing and complex civilization was timber. In Europe, also, intensive agricultural practices coupled with heavy animal manuring made the cleared land even more fertile for a while than the forest soils it replaced.

The speed of deforestation matched the progress, or otherwise, of civilization. When the Roman Empire disintegrated after the fourth century AD, forests reclaimed the abandoned fields. But after AD 1500 timber shortages in Europe were becoming acute, and once again the forests were plundered. As the forested areas shrunk remorselessly, wood became more valuable, and the ruling classes even attempted to preserve the woodlands by serving prohibitions on the activities of the peasantry.

In Scotland the forests were cut down to serve as fuel for ironworks. The English needed much timber for their war galleons, and New Zealand lost 15 million acres of woodlands shortly after English migrants arrived there. The Pilgrim Fathers and their descendants soon turned America's early forest empire into a memory. In the late nineteenth century, because of the dominance of wood as building material, as much as 85 per cent of the forests of New England were being chopped down.

China, however, has suffered even more than Europe from wanton tree felling. At the time of the Shang Dynasty China was covered with forest. Then trees were cut down as the need for more agricultural land became pressing. The eroded loess, washed down as silt, made the lower plains fertile. But it also eroded that soil from elsewhere – transporting some 2,500 tons of the stuff annually. Soon it raised river beds and increased the frequency of flooding, both from the rivers and from the storm waters rushing down denuded hillsides.

Chinese scientists are now warning that further disastrous floods are in the offing unless more effective action is taken to replant forests on the upper reaches of the Yangtze and Hwang Ho. The Yangtze floods of the summer of 1981, costing some £700 million, came dangerously close to inundating the big cities of Wuhan and Shanghai. It was only the low water level in the Donting Lake, which drains much of the river's water, which averted the worst of the damage.

Dr Hu Qingjun, a river control expert, has pointed out that serious, life-destroying floods occurred in the Sichuan province in the nineteenth century when extravagant rulers plundered the hillsides for timber to build huge wooden pillars for their palaces. Perhaps Dr Qingjun was trying to make a political statement. What he did not mention was that the 'grain first' agricultural policy practised under Mao Tse-Tung is largely to blame for inappropriate terracing and other landscaping methods which have increased erosion. It has been a costly exercise, as there has been little extra grain to show for it.

The truth is, as we have seen, that China's flood problem, largely the result of pressures of population, dates back to the second millennium before Christ. Still, it cannot be denied that deforestation in China has intensified in the past 30 years. Indeed, there is rabid competition among numerous logging teams working under different local authorities.

The Vanishing Flood Barriers

Serious consequences can arise when trees, nature's own flood barriers, are felled. Ultimately the soil beneath the vanishing trees undergoes irreversible biological changes. As we have seen, the dense vegetation of some floodplains has a spongelike quality about them. In other areas rainwater falls directly onto the soil. How much sinks into the ground and how much runs off depends, of course, on the porosity of that soil.

The dark humus held in place by the roots of plants is protected by their tops from the harshness of the elements. And when plants die their roots rot in place. The dead leaves of grass form a mulch or blanket which is the ideal condition for

Throughout history Asia has suffered particularly from very serious flooding. This rare print shows a Malayan town under water, *c.* 1910.

139

water to land on. This enables the water to percolate down to the soil which absorbs it, without sinking in too quickly and saturating the ground. Eventually the moisture is returned to the hydrological cycle through trans-evaporation.

The worst that can happen, before the flood-stage is reached, is that the area will become a marshy quagmire. Hence the presence of vegetation is a variable that is independent of the porosity of the soil it grows on, and in the absence of vegetation excess rainwater will have to run off somewhere. Not only that, but the ground would fast become impermeable as soon as the binding action of plant roots is destroyed and the soft topsoil removed through erosion. What few minerals that may be left in the subsoil are washed deep down, where they become lost to forest flora. And the sun, of course, bakes the ground hard. Thus the probability, indeed certainty, of more floods.

However it is the washing away of topsoil by land newly denuded of greenery that creates a vicious ecological circle. The hydrosphere and the geosphere are constantly in a war of attrition: the sea, the rain and rivers all eat away at the land regardless of what Man is doing. The porosity of the earth is eroded both by wind and by overspilling headwaters.

A severe flood can easily wash away the topsoil of even agricultural or arable land, thus making each succeeding deluge that much more damaging. One experiment in the Amazon showed that 85 inches of rain a year removes less than half a ton of soil per acre from a forested area sloping at 15 degrees. But when an area of forest is removed, even though it was on a level terrace, *45* tons of soil is washed away.

Furthermore, in the interior of large continents the rainfall is much less than on the coast, and the vegetation is thus much sparser. Yet when the rains come, they fall in torrents but encounter the least resistance. So the ground floods easily. The rain carves great gullies, while in the dry season the now unfettered wind blows the dusty surface away, often at the rate of several inches a year.

In the Western Mississippi Valley the vegetation is short grass and shrub, and the rainfall is about 15 inches a year on a hinterland nearly 1,500 feet above sea level. So the floods fall from a greater elevation, thus adding to its weight-uprooting properties all the way on its southbound journey to Oklahoma.

Thus it is imperative that the right sort of greenery is planted. It was estimated that the 1947 Missouri floods tore away more than 115 million tons of rich loam topsoil, which had hitherto made the state of Iowa one of the greatest agricultural areas of the world. In his famous early doomwatch book, *Road to Survival*, published two years after the Missouri floods, William Vogt said that the wrong sort of crops had been planted – corn and soyabeans in row-tills – and were more liable to be eroded than forest or grassland.

Slashing, Burning and Nibbling

Man's continuing intervention in the ecological cycle has recently greatly speeded up the process of earth-balding and soil eroding. Many early civiliza-tions, as we saw in Chapter Four – the Sumerians and the Indus Valley dwellers – had moist and fertile lands in abundance. Climatic change dried them out somewhat. But a decrease in rainfall levels would certainly not have accounted for the rapid rise in aridity in the now inaptly named 'Fertile Crescent' if it were not for the over-exploitation of soil fertility by nomads, peasants and farmers.

There can be no doubt about the pernicious effect of humanity on the world's plant life. The ravaging of earth has been going on for some thousands of years as Man burnt trees and undergrowth for fire, light and cooking, and to improve mobility when hunting, and to increase grassland for pasturage. In many parts of medieval Europe over-grazing by sheep and goats has taken its obvious toll. Goats are highly destructive – they can insidiously reduce forests by nibbling away at the bark and sprouts of trees. They were certainly a critical factor in reducing fifteenth and sixteenth-century Spanish agricultural land to the state it is now.

The floods of the Arno have been a regular occurrence since the fourteenth century when the woodlands around Florence were converted to pastureland. And the over-grazing by farm animals ultimately turned the ground into baked clay. The earliest Arno flood was in 1333 when 300 were drowned. After that re-afforestation projects were urged, but ignored. Since then there has been a major flood every 100 years or so, and no reafforestation is even now taking place. Indeed no serious attempt has been made to dredge or deepen the existing riverbed, and six feet of silt still remain from the 1966 flood.

But it was the last war that saw a marked speeding up of earth ravaging, tree lopping, activities. Goats, firewood scavengers and the British troops in North Africa and the Far East put an end to the few remaining woods still standing and left an exposed topsoil that soon became pitifully thin.

In his 1970 *Doomsday Book*, Gordon Rattray Taylor tells how, after the war, the Japanese were urged by the Americans to improve their agricultural output by converting forests to farmland. But the Japanese, having learnt from the follies of the Meiji period when deforestation had caused disastrous floods, had already imposed strict forest conservation laws. But in 1945 they were virtually obliged to fell, with, as Taylor says, the same results: floods and erosion. Before long the Japanese re-enacted their forest protection code.

Bengal has also suffered since the war from tree clearance programmes in the Himalayan foothills which are geologically new and disintegrate easily. This has naturally had repercussions hundreds of miles away, to add to the flooding risk brought about by cyclones. In India more than 6,000 million tons of topsoil – ten tons for every person in the country – slips away each year.

As the soil is remorselessly eroded year by year by man and his animals so the world's peasant farmers advance deeper into the forest, slashing and burning anew, and the cycle is repeated. In Haiti slash-and-burn farmers have succeeded in converting most of the upland areas into rocky plateaux. Across sub-Saharan Africa millions of domestic fires have chewed barren gaps in the jungle. The Sahara Desert, in the meantime, expands by 100,000 hectares every year, made worse in recent years by severe drought conditions. As the trees become literally thin on the ground African women become more and more enervated as their working week is spent hunting further afield for firewood.

But this constant ravaging means that deforested areas are not allowed to

Brazil – State of Amazonas. An aerial view of jungle country above the Caracarahy Rapids. The slashing and burning of the world's forests worsens the flooding threat, and could ultimately affect the global climate.

recover naturally. The increasing shortage of firewood – a principal fuel for three-quarters of humanity – forces the exhausted women to hunt for crop wastes and dried up excrement – anything that will burn – for cooking and heating, thus denying whatever good soil is remaining the necessary nutrients for continued survival.

Even now, where a more repectful attitude to horticulture is combined with a scientific approach, there is still much to suggest that the soil is being inexorably overworked by the use of fertilizers. Tree planting has been carried out only spasmodically, such as in Massachussetts. On the whole it has been grossly deficient, and done mostly with the faster-growing softwoods first which can disturb the ground's nutrient balances.

But the worst victim, by far, is South America. There the forests suffer much more from commercial activities than from peasant farmers. The twentieth-century demand for timber is now largely to feed the newsprint mills. The US National Forest Reserve believes demand for timber and newsprint will double again by the year 2010.

Prices of pulpwood in the meantime have been rocketing. Two-thirds of Latin America's forests have gone, or have been seriously diminished. Beef raising has grown apace, spurred on by rising prices in the developed world because of what is known as the 'hamburger connection'. So huge cattle ranches require more pastoral land, so again the forests – in a land seemingly everlastingly endowed with trees – take second place. In one case a multinational corporation was said to have burnt down a million acres of forest in the Amazon Basin to make way for a cattle ranch. The fire was said to be so big it was reported by a weather satellite as an impending volcanic eruption.

Other areas of the globe suffer from human rapacity. The West has an insatiable appetite for veneers and rare hardwoods – to make pianos, mahogany coffee-tables and teak floors, for example. The search for these marketable hardwoods is concentrated in south-east Asia. Tree felling has been so successfully pursued that virtully all the lowland forests in Malaysia and the Philippines will be depleted by 1990. Thailand is already forced into the position of having to *import* wood products. Logging concessions for Japanese multi-nationals (Japan is a major consumer of hardwood) have dried up. Japan, seeing the light, had to appeal to UNCTAD for a new deal for producers and consumers.

Deforestation and the Moisture Balance

Cutting down the forests inexorably leads to more floods. But, paradoxically, it also reduces rainfall in many local areas. The reason for this is that more solar heat is reflected from the now bare soil, which brings about changes in air circulation and weather patterns. The key principle involved is the reduction in transpiration (the exuding of water vapour) from the diminishing leaves of the forest.

The British Institute of Hydrology, in studies of English woodlands, have shown that the forest uses more water than other vegetation. Firstly, it has a lower albedo, or reflectivity, rate of 17 per cent, whereas grassland has 26 per cent. This lower albedo means there is more energy available for evaporation, i.e. the canopy of leaves is that tiny bit hotter than the surrounding terrain, so speeding up the evaporation of rainfall. Secondly, the institute believes that the deep canopy enables a greater mixing of air to pass through the forest, again speeding up evaporation. And as we have already seen that rain is produced by heat plus evaporation, it follows that the less forest there is the less rainfall there is.

However, the argument is not so simple as this. The question of an increase or decrease in rainfall depends largely on whether the rain falling onto a forest is the product of recycled water vapour, or whether it is a primary vapour arising from the oceans. Eneas Salati, of the National Institute for Amazonian Research in Brazil, was able to prove that only half the Amazonian rainfall was recycled vapour.

Marajo Island in the mouth of the Amazon River provides an excellent test of the cloud cover thesis. For the western half of the island is virtually all forest, while the eastern savannah is bare. The forested half is nearly always cloud-covered, and rain pours down daily. The eastern half gets no cloud and hence no rain.

Hence, deforestation in local terms reduces water vapour. But the shifting winds of the weather machine will still transfer moist air from other forested and vegetated areas, so that things get evened out, say over a deforested region of a few hectares. Over a larger area, however, there is more likely to be a bigger drop in air moisture. And if this drier air is not dispersed by the prevailing winds, it could dry out a cleared area and produce more erosion. This in turn will gradually affect the remaining forest. And as more of the forest is hacked down by Man, this drying out phenomenon will get worse.

The crucial question is how deforestation will affect *global* rainfall. Here we must make a guess at whether earth's cloud cover will increase or decrease. Moist air movements from ocean to land, from the northern hemisphere to the southern, are constantly at work, largely overriding purely local effects. We must bear in mind complicated variables such as the rise in convection heat from urban areas and natural climatic changes.

The Proliferation of Heat Islands

Hence no discussion of rainfall levels would be complete without an acknow-ledgement of a vital factor that may override all others: the 'heat island' effect.

Again we must remind ourselves of the warmth + moisture = rain equation. Before mankind arrived on the scene these rain-inducing characteristics arose naturally from the biosphere – the earth, its biological life and its atmosphere. But Man, by himself, is an energy converting machine. And according to a basic principle of science we call the Second Law of Thermodynamics this crucial fact means that earth should be getting warmer because energy is being used by Man and his machines on an ever-increasing scale. All of this energy is eventually, and inexorably, degraded into heat. Just sitting quietly, a human being gives off 450 British Thermal Units (BTUs) of heat per hour, which is enough to warm one pound of water to 1°F. A large family saloon driven at 70 mph gives off 750,000 BTUs. In 1925 total energy consumption and conversion had already reached 44 quadrillion BTUs, and, growing at a rate of 5 per cent annually since then, world energy use leapt to 345 quadrillion BTUs in the 1980s.

In addition we have built ourselves massive heat and moisture emitting cities across a great part of the world's land surface. And they all generate more heat than they receive from the sun: New York emits seven times as much. In a sense we are incubating the world. And although the population explosion has for years been deplored for 'environmental' reasons, the serious climatic impact brought about by the sheer size of the human family has only recently been appreciated.

In 1980 the world's population reached 4,415 million, of which 3,284 million were in the developing countries. And it is in the so-called Third World that cities are growing fastest. Only 300 years ago the world carried about 500 million people, and 1,500 million have been added since 1930. The total mass of humanity now weighs about 180 million tonnes, more than half the total mass of the earth.

By 1985, as the exodus from rural areas continues, Tokyo will house 25.2 million inhabitants, Mexico City 18 million, Bombay 12.1 million and Los Angeles 13.7 million. By 1985 London will have been overtaken in size by Seoul. Latin America underwent an 80 per cent increase in population between 1950 and 1980 when the figure reached 380 million. But this still cannot compare with the startling growth rate of the Philippines, which suggests that in a century from now its population will not only equal today's Latin American total, but those of North America, Africa and the USSR as well.

We can be reasonably certain that all these growing cities must warm the earth. For one thing, the materials of which urban areas are made – stone, concrete, asphalt and brick – stores heat more rapidly than natural terrain. Cities also absorb more sunlight because of the prevalence of steep non-reflective walls and the jagged profile of the skyline.

Of course, dry convection itself does not produce precipitation. But invariably there is sufficient moisture in the atmosphere to ultimately cause rain over the hottest, driest city summer days. It just means that the heat will have to rise further into the atmosphere before a cloud will form. Indeed, condensation and humidity arising from city air now more than compensates from any moisture loss due to sewer run-off. Certainly it has been proved that thunderstorms over cities, especially as warm sultry summer days draw to a close, are much more prevalent than over open moorland.

So the 'heat island' effect grows more pronounced as in each decade more arable land falls victim to the bulldozer. Ultimately, as this extra heat drifts futher afield, complicated weather anomalies arise. Climatologist Stephen Schneider, in his book *Genesis Strategy*, says there is good statistical evidence that rainfall up to 50 miles downwind of industrial areas has grown by as much as 15 per cent because of the extra heat and dust released into the air.

145

The Greenhouse Effect

But the chief worry of weather experts today arises from the well-known greenhouse effect. Over the years carbon dioxide (CO^2) is discharged into the air in increasing quantities. The theory holds that the gas allows the sun's shorter unltraviolet wavelengths and visible radiation to warm the earth. But at the same time it absorbs the longer infra-red wavelengths, the surface tries to radiate back into the atmosphere. So the heat builds up, as it does in a greenhouse.

The effect is largely a postwar phenomenon because whereas pollution is controllable (and has largely been abated) carbon dioxide is made by Man's most basic economic activity: the production of energy itself.

A residual amount of carbon dioxide, of course, exists naturally in the atmosphere. Via photosynthesis – a complicated process utilising sunlight – vegetation and foliage converts the gas into carbons, and builds complex molecules that make up the tissue of all plants. When living matter dies off it gives off CO^2 by the splitting up of its hydrocarbon molecules.

But today there is far more CO^2 in the air than exists naturally because, in a sense, Man is unlocking the past. Millions of years ago plants became trapped in the bowels of the earth, when the climate was tropical, to form today's giant reserves of fossil fuel such as coal, oil and natural gas. It is the consumption of these hydrocarbon substances that has fuelled, in the strictest sense, the Industrial Revolution. The CO^2 stored in the earth is now being returned to the atmosphere, releasing about 20 per cent more than would occur naturally without Man's intervention.

Indeed, in one year as much coal can be extracted from the earth as took 400,000 years of dying trees to leave behind. In a mere decade – from 1960 to 1970 – coal production rose by a quarter, and natural gas by 40 per cent, whereas the use of electricity shot up by 60 per cent. Hence the seriousness of the Amazonian forest burning – vast amounts of carbon dioxide are discharged into the atmosphere within a few years, although the trees themselves are the end-result of millions of years of growth. The process is self-generating; as deforestation increases aridity, so bogs and marshlands dry up releasing still more CO^2. In the language of cybernetics there is a 'positive feedback'. (A lot of this carbon dioxide might be utilized by an increase in herbaceous vegetation which has a higher rate of photosynthesis, and thus is able to remove the gas quicker).

Estimates of how much fossil fuel remains vary widely, but a consensus view is that only 7 trillion tons of fossil carbon remain in the earth. And at the present rate of consumption all those tons of carbon might be used up in less than one century.

Petrol burning vehicles also contribute about half of some air pollutants – mainly carbon monoxide (CO) – observed in big cities. US roads cover nearly 1 per cent of the entire land space, which is not as big a change of land as that due to forests making way for agriculture. But this percentage is bound to grow. According to the OECD the number of vehicles in the world rose from 100 million units in 1960 to 200 million in 1970, and reached a rough estimate of

300 million in 1980. Exhaust products from jet airliners – which include water vapour – are deposited high in the troposphere and lower atmosphere. According to the Massachussetts Institute of Technology, who have published findings on 'inadvertent climate modification', jet traffic has already caused a small increase in cirrus cloudiness in heavily travelled areas. This, the institute believes, must have an effect on the earth's heat balance.

Research by scientists in the carbon problem began in 1958 as soon as regular and accurate measurements of particulates in the atmosphere could be made. At that time it was estimated that the amount of pre-industrial CO_2 in the air was somewhere between 275 and 285 parts per million (ppm). Since the late eighteenth century 400,000 million extra tons of carbon-based pollutants have been pumped into the sky – CO, CO_2 and hydrogen (in the form of water vapour, H_2O) and various sulphates. In 1958 it rose to 315 ppm, and in 1977 the average value of CO_2 in the air was 330 ppm. Over half this increase had been put there in the last thirty years. Currently the rate of emission is 40,000 tons a minute.

As carbon dioxide emissions rise, then, surface temperatures warm. The earth is probably 1°C hotter than it was perhaps a century ago, but quickly getting warmer. True, this is insignificant compared to the climatic changes that the global weather machine can bring about, including countervailing trends towards a cooling.

And we must bear in mind the important role the oceans play in the CO_2 and heat balance syndromes. Climatologists have programmed computers to help analyze the earth's heat balance as it is distributed across regions and oceans, and then inserted mathematical imputs to represent the putative extra amount of annual carbon gases. There is an exchange of an estimated 113 billion tons of carbon a year between the land and the atmosphere, and about 90 billion tons between the oceans and the atmosphere.

But the oceans are still an unknown variable. Scientists from Scripps Oceanic Institute aboard the US research ship, *The Knorr*, were in the late autumn of 1983 making good progress in determining how the Atlantic Ocean's surface reacts with the atmosphere. They were anxious to test theories, such as that held by Minze Stuiver of the University of Washington, that the oceans are more efficient at absorbing CO_2 than previously thought, and if so whether this process could be accelerated and whether saturation point would ever be reached. They are particularly interested in the icy flow of water from the Poles into the warmer regions, since colder water holds more CO_2. The oceans, in fact, have a mighty knock-on effect that is hard to beat: as global temperatures rise so ocean evaporation causes greater humidity and thus reinforces the warming effect. This is because H_2O molecules are even more effective infra-red absorbers than CO_2 molecules.

Other intriguing complications can be observed. Greater evaporation produces more cloud cover that will block out sunlight and hence cause a local cooling. It is for this reason that greater earth warming – because of the likelihood of more precipitation – is more likely to spell more floods rather than an increase in terrestrial aridity.

147

The seafarer's first glimpse of Antarctica. Three miles thick and larger than Europe, this massive continent may shortly melt and cause widespread flooding around the world.

The Return of the Big Melt

But there is another reason why a warming would increase the likelihood of floods, and it has received much attention by doomsday writers in recent years. By the turn of the century the atmosphere will contain up to 20 per cent more CO_2 than it did 100 years ago. This is the conclusion of one of many scientists, the Soviet climatologist Mikhail Budyko. And William Kellog, of the National Centre for Atmospheric Research, believes that by the middle of the next century, when we may have twice as much of the gas as we did in 1900, the global temperature could rise by an average of 2 or 3 °C. In the polar regions, however – and this is the worrying aspect – the temperature could become 5 to 10 degrees warmer. It could dramatically alter the whole character of the general circulation of weather patterns.

Scientists are concerned that both, or at least one, of the polar caps will melt and that the earth's 9 million cubic miles of ice would melt. But the period involved would be remarkably short compared with the rest of geologic time. All of the world's great cities are really ports, and they would all be at risk even in the very early stages of the melt.

The Greenland ice cap itself contains 620,000 cubic miles of ice. If that and the lesser ice sheets on other polar islands were to melt ocean levels would rise by at least a few feet. But the real danger arises when the Antarctic melts as well. The rising tide would spill over about two million square miles of low-lying land, up to a speculative depth of anywhere between 2 and 200 feet. Certainly by the *end* of the melt most coastal cities would be drowned. The seas, by the time all or most of the polar caps had dissolved, could even, it has been suggested, reach up to the twentieth floor of the Empire State Building.

Leningrad, Venice, Rio de Janeiro and Bangkok would also be seriously affected. Around the globe large areas of lowland would be submerged – both sides of the Flemish Bight, for example, and Florida. More than two million people in London would be at risk, as the flood waters reached Piccadilly. The House of Commons would be inundated. In their book *Earthshock*, geologists Basil Booth and Frank Fitch assert that even if the melt took 200 years, London would flood during the first *six* years. And even if the ice took 1,000 years to dissolve, London would flood in 28 years, 'perhaps less'.

By then, they believe, the Netherlands, the Po Valley in northern Italy, and much of northern Europe to the shores of northern Russia 'would all sink below the advancing tide of rising water'. Parts of the Middle East, Bengal, China and large regions of South America would be particularly badly hit. Northern Canada and the Mississippi Valley would be awash.

Some experts, however, say that this doomwatch scenario is unlikely. They say that the ice sheets have survived all the interglacial periods of the past. Some even deny a melting could take place within the temperature ranges cited – up to 10 °F. But the arithmetical logic of climatic change cannot be ignored, especially if one believes that earlier Big Melts were triggered by increases in solar radiation spread over longer time spans. Even a mild rise in atmospheric temperatures can cause the oceans to swell. For example, between 1930 and 1948, which climatologists generally agree was a warmer period than the present, correlated measurements from geophysical laboratories around the world suggested that the global seas rose six inches, estimated to be four times the average rate of rise for the last 3,000 years. And in the 1920s the rate of increase was even higher than this.

Another disturbing pointer to the plausibility of the Big Melt thesis is the instability of that region where 85 per cent of the world's ice packs are stored – the Antarctic. The region seems to be hair-triggered to cause sudden ice melts and freezes. The eastern Antarctic is much bigger than the west, and is grounded on land lying above the sea level. And it is these ice sheets that are thought to be responsible for past worldwide glaciations. Glaciologists from Arizona State University talk of an ice surge and a melt surge. In the former, accumulated snow and ice abruptly gain momentum when the ice sheets exceed a critical thickness of about three miles. The latter phenomenon is where an opposite minimum ice thickness – given the right atmospheric conditions – causes an abrupt melt.

But the Ross Ice Shelf, in the western Antarctic, is an equally dangerous climatic time bomb. This floating slab of ice, extending some thousands of miles, is also unstable, as it is only tentatively hinged onto islands and is well below sea level. It in fact acts as a buttress to the land-based ice in the Antarctic, holding it in place. If the climate warms by a few degrees, these supporting shelves could break up, thus releasing enormous geotectonic pressures.

As the ice breaks away it bounces up vertically as it is suddenly freed from its terrestrial tether, displacing millions of gallons of water to cause tidal waves in the southern hemisphere. However, as the melting of floating ice does not actually increase the *volume* of water, it is the east Antarctic that remains the greater threat.

The concern about the greenhouse effect was heightened in October 1983 when the US government's Environmental Protection Agency published an official study which said that the warming will come sooner than expected. The study says that current estimates forecast a 3.6 °F rise by 2040 and a 9 °F increase by 2100, and could lead to 'catastrophic' social and economic distruption.

Mr John Hoffman, director of EPA's strategic studies staff, says that the warming of the world is 'neither trivial nor a long-range problem'. The heat-up of the poles would give Charleston, South Carolina, a tide of as much as seven feet higher 120 years from now. It is too late, say the EPA, to do much about it, even if a total ban on fossil fuels were imposed now. While other government studies have warned of the warming, this latest report is the first to state with certainty that it *will* occur no matter what we do about it!

The world's weather will also be adversely affected. 'Temperature increases are likely to be associated', said the report, 'by dramatic changes in precipitation and storm patterns.' New York will become semi-tropical, and the effect on American agriculture will be felt around 2000.

To underline the EPA's stark message, a simultaneous report by the National Academy of Sciences confirms that the melting will cause serious geophysical problems, and came up with the latest estimate of the CO_2 concentration to date – a doubling up of present concentrations by the third quarter of the next century. But the most alarming statement on the greenhouse effect has come from NASA's Brian Toon, who has been studying the effect on the 900 °F surface of Venus. He concludes: 'We're on the ragged edge of convincingly demonstrating that it's happening on earth as well'.

Not all climatologists are convinced, however. And some are optimistic about the effects of any warming. There should, they believe, be adequate time for coastal dwellers to be relocated. And there would be the added advantage of a pleasanter climate, with a larger percentage of ice-free habitable land that could be exploited by an enlarged population. And in northern climes the growing season would become longer.

Chapter Eleven

OUR TEETERING CLIMATE

Since the early 1970s the world's weather has been little short of appalling. Almost every month now the news media draws our attention to freak weather – floods, droughts, ice, freezing 'cold snaps' and heat waves.

Countries seem to have been taking it in turns to suffer these extremes. While Africa and Australia have been parched, America has suffered from an excess of moisture rather than the lack of it, with catastrophic ice-melt floods and destructive coastal storms.

In January 1985 the weather was so cold in sub-zero Washington that the presidential inauguration ceremony – for the first time in history – could not be held out of doors. February in Europe was the coldest in living memory. Most of the Baltic Sea froze for the first time since the war, making it possible for Poles to stroll over to the Swedish island of Bornholm.

However, the extraordinary weather of the first half of the eighties was its thoroughly schizoid character. Most continents – Africa, India and South America – were suffering from both floods and drought at the same time; a mere shift of 50 miles or so in longitude determining whether one or other phenomenon would arise. Many other parts in both the northern and southern hemisphere from Polynesia to the Americas had been hit by floods. Everywhere storms had turned parched scrubland into swamps, and had sent rivers of raw sewage through Latin America's teeming cities.

The African drought, exemplified on Western television screens by the emaciated refugees from Ethiopia, was blamed for up-ending South Africa's whole economic system and fragile status-quo. Less fortunate Latin American states, already struggling with massive foreign debts, were threatened with ruination while hundreds of deaths had already occurred in 1983 as villages were inundated with giant mudslides. There were floods in Paraguay and Uruguay. In July 1983 at least 30 people died in Rio de Janeiro as torrential rains pounded the southern Brazilian states of Rio Grande de Sul and Santa Catarina, where 200,000 were forced to leave their homes. Heavy rains had been falling on the highlands and coast of Ecuador and northern Peru since late 1982, causing much damage to crops and houses. In northern Peru a wall of mud and debris 50 feet high and more than 100 feet wide had surged down a dry river bed to claim 60 lives.

And yet, typically, southern Peru and parts of Bolivia had been suffering from one of the worst droughts in living memory. As a result of both floods and drought more than 230 Peruvians had died. According to the Peruvian government at least £500 million worth of damage had been wreaked during the period September 1982 to June 1983. Torrential and incessant rains in the 'desert city' of Piura broke up roads and undermined flimsy buildings, menacing the lives of its 200,000 inhabitants, many of whom had never known rain.

The Ecuadorean authorities had already declared the capital city, Quito – 9,000 feet up in the Andes – as a disaster area after fierce rains had caused dozens of mudslides. More than 100 were said to have died at Chunchi, in the Chimborazo province, when mud and rocks fell onto lorries and buses lining the road below. For months in the port of Guayaquil, the country's biggest city, the two million inhabitants had endured their streets being polluted with streams of water covered with vivid green slime, after the skies had discharged 15 times its normal May rainfall. Along the full length of the Peruvian coast virtually every bridge on the vital Panamerican Highway had been washed away.

Central America, to compound its crucial political problems, was in turmoil. El Salvador, still reeling from the effects of official and unofficial death squads, had already been the victim of nature's own terror when 500 were killed and 25,000 made homeless from the previous year's floods.

Things were little better in Australia. Floods hit the states of New South Wales and Queensland after brutal rain that fell for more than a month after years of drought. Estimates put the direct losses at about £150 million, with the final sheep loss at 400,000, many dying when they could not stand under the weight of up to six gallons of water soaking their fleeces. The Ministry of Primary Industry declared that there had been a severe loss of topsoil with substantial soil erosion. At the same time it was raining heavily in California, and floods had been reported in that familiar floodplain, the Mississippi area. In the deep south 25,000 folk had to flee their homes, and a further one million were isolated.

The whole of the Pacific coastal region, north, south, east and west, was buffetted by cyclonic weather. In the western Pacific flood waters from China's Yangtze River were fast approaching the levels of the 1954 disaster, threatening the provinces of Hubei, Hanan, Sichuan (which has precipitous mountain ranges) and Jiangxi, which have a total population of 227 million. In western Japan more than 120 people drowned in landslides and floodings triggered by heavy rain. And in the north-west a peculiar Pacific aberration to the flood problem – the tsunamis – had whipped up waves 30 feet high in places and caused the deaths of more than 100 people, including at least 12 children picnicking on a beach.

In India the equatorial currents that had plunged through Indonesia had brought about the deaths of at least 2,000 people in the state of Gujarat alone following just three days of monsoon rain. The Press Trust of India news agency, reporting from the worst hit Junagadh district, described how waves of murky water had left a trail of death and devastation, with the corpses left dangling in trees. In Wanthali hardly a house or shop had escaped the floods, and an official said an entire family of fourteen had been swept away.

A New Storm Age?

Since the 1960s the world's weather has become increasingly erratic. During the early 1970s the continental USA suffered severe floods which destroyed billions of dollars worth of property in the Mississippi Basin, the Great Lakes region, Pennsylvania and New Jersey. The 1975/6 winter was severe, and rain and floods, the worst for a century, struck large parts of Washington State. Record snows fell in California, thus ending the state's severest drought, also the century's worst.

In January 1976 German meteorologists were predicting that the north-west coast of Europe could be facing a new 'storm age' when the threat of flooding would loom menacingly large, and even disrupt North Sea oil production. As it happened, there were two very severe coastal storms along the English east coast in January 1978 and 1979. Whiplash storms returned to Europe in 1983. While many in northern parts were revelling in an unprecedented heatwave, people in other parts were the victims of raging floods. Nearly 50 died in Spain and the Basque Country as surging water reached a depth of over 20 feet in some places, demolishing buildings, submerging towns, sweeping away hundreds of cars and blocking tunnels. Monsoon flooding affected 22 of Thailand's 73 provinces and killed 11 people, destroyed roads and left 800 families homeless.

The previous year, 1982, had also been an appalling one for weather. Out of eleven serious events in 1982 five were highly damaging floods, and two more (in Italy and Tonga) were moisture based disasters, while there were two instances of a flood/drought interface. Storms had ravaged southern France and Spain for three days, and the Costa Blanca flood took the lives of nearly 100 people.

Was this the trend for the future? Apparently so, thought the Chinese in 1983 when the Yangtse rose to disturbing heights, and memories were stirred of the floods a year earlier when 1,358 villagers had died in Shanxi Province, and yet another 400 in Guangdon in 1981. The weather experts were not slow to observe that these floods had also preceded serious droughts in the Hobei and Liaoning Provinces. The patchwork schemata was even more marked in Thailand, where one village suffered drought while its neighbour only 30 miles away was inundated.

Floods in Nagasaki, Japan, in 1982 also took the lives of more than 300 in that city's worst disaster since World War II. Similar floods and hail had ruined arable land in the Soviet Union and destroyed livestock in a country notoriously prone to food shortages for either institutional or climatic reasons. Crops, too, in southern Italy and Sicily were extensively damaged by heat and hailstones. Palermo experienced 92 straight days of near 100 °F – a record.

The US, it seemed, had not experienced a normal spring for many years, with hail in 1982 costing Texas cotton farmers $2.2 billion. In Northern Indiana, Ohio and South Michigan, rain-swollen rivers resulted in the worst flooding there since 1913, with more than $20 million worth of damage done in the city of Fort Wayne. Later, in December, heavy rainfall totalling more than 20 inches in three days sent swollen rivers swirling over their banks. Again, in the first week of February 1983 exceptional rainfall and strong seas created nightmarish conditions for coastal dwellers. A storm in the Gulf of Alaska lumbered southward towards California, gaining in force and intensity. By the time it struck the coast of America's most populous state it hurled driving rains and 15 foot waves onto beachside restaurants, shattered the surface of highways and sent mud cascading down hillsides. Four counties, including Los Angeles and San Diego, were declared disaster areas. The famous Pacific Coast Highway through Big Sur was split in 30 places and was not repaired until autumn.

Was El Nino to Blame?

By mid July 1983 a new theory to account for the world's weather was being discussed by meteorologists. Experts at the US National Oceanic and Atmospheric Administration had noticed that the surface of the Pacific Ocean had warmed up. A great wave of warm water was discovered surging its way to the coast of Latin America. They were already aware, of course, that the Pacific, being the world's largest ocean, was the major component in the global heat engine.

But this additional warming was something new. Mid surface temperatures had been raised by as much as 11 °F, while the sea level was alarmingly swollen by some seven inches or more. The weathermen began borrowing a name often used by South American fishermen to explain these atmospheric disturbances – 'El Nino'. Normally occurring around Christmas time, in 1982 they had begun much earlier, and were more virulent in their effect.

This meant that when the air pressure was relaxed, warm surface water that would normally be pushing anti-clockwise round the so-called Peru Current was surging back instead towards North and South America. Accompanied by strong air currents and torrential rain, it set off a chain of unpredictable disturbances reaching far into the jet-stream currents snaking 40,000 feet above the earth. Simply put, Pacific weather had been reversed, bringing drought instead of humidity to southern India and Australia, and floods to California.

Or was it El Chichon?

But the El Nino theory, purporting to explain new and probably temporary weather disturbances, had other serious contenders. One of these was the volcanic dust theory, and has experienced a revival of interest among climatologists. The theory dictates that volcanoes erupt with such force that clouds of ash actually darken the sky across the globe. Some scientists believe that volcanic dust may linger in the air from 2 to 12 years, and be 30 times as effective in blocking incoming radiation than retaining heat as carbon dioxide does. Whereas man-made activity tends to pollute the lower layers, volcanic dust can rise much higher, and can cause some 70 per cent of earth's temperature variations.

A great deal of interest was stimulated by El Chichon's 1981 eruption which broke a 5,000 year silence. Scientist with NASA's Aerosol Climatic Effects Programme are using sophisticated ground-based lasers to track the height of erupted materials. High-flying U2 aircraft are used to collect and sample what were first thought to be dust particles only to find that most of the spewed material was sulfur dioxide gas. What, they discovered, was blocking out the sunlight was not ash particles but a curious blend of water vapour droplets fused with acid. And unlike the heavier ash they could remain suspended in the atmosphere for years.

For this reason the Mount St Helens eruption in the US in May 1980 was not thought later to have had much impact on the world's weather, since satellite and aircraft readings later proved it lacked the necessary sulfur. In addition, although Mount St Helens erupted with more fury, the barrel of the cannon was pointed laterally, whereas El Chichon was pointed straight up, forcing the dust and gas to travel further.

Still, the cause and effect connection was poorly understood. The theory that the aberrant weather of 1982 was due to El Chichon was challenged by researchers at the Climatic Research Unit at the University of East Anglia, arguing that the time lag was too great. It would have been more plausible to blame the *previous* June's low temperatures in Britain on the effect instead. American scientists, on the other hand, were insisting that the effects of El Chichon could last for two or three years, enough to chill America and Europe by 1 °F overall. By January 1983 some were already attributing the hard freeze in Delhi and New Orleans to accumulating dust particles in the upper atmosphere.

The classic example of the link between volcanoes and weather was first established after the Indonesian volcano, Tambori, erupted massively in 1815. The summer of 1816 simply never occurred. The last century, culminating in Krakatoa, was dominated by vulcanism. Then there was a quiet period until after the World War II, when eruptions started again in earnest with the more notable volcanoes being Mount Agung, Galunggung, Mount St Helens, and El Chichon.

Now the connection between volcanoes and weather is all but positively established. Professor Hubert Lamb has developed a Dust Veil Index, and says that many of the wettest summers in the western world have occurred when the Index – a handy yardstick taking Krakatoa as a baseline – shows a higher reading than usual.

John Gribbin, the atrophysicist and science writer, in his recent *Beyond the Jupiter Effect*, believes that the eruptions of the 1980s are harbingers of a new pattern – 'or rather a return to normal' – a theme he has reiterated in many books and papers. Even in September 1982 Gribbin was reporting, in a *Guardian* article, that there was a 5 per cent reduction in the amount of heat arriving from the sun, and quoted NASA physicists who predicted a cooling of about 2 °C 'for months or years'.

The 'Blocking Highs'

While the scientists theorized, the bad weather continued. In 1983 European skies wept endlessly. A great many weather records were broken while the Met men held their judgement or put it down to the natural vagaries of climate, or just 'bad luck'.

The wine-growing Beaujolais district near Lyons had its worst floods in 40 years, following nearly 10 months of rain. All the major river systems of Europe lapped over their banks, some up to four times: the Seine, Garonne, Loire, Rhône, Rhine, Moselle, Neckar and Saar. The Rhine, in fact, swollen by torrential rain of long duration, roared in a brown torrent through the old city centre of Cologne, up to five feet deep, and there were fears that historic buildings might collapse. Low lying areas of Koblenz and Bonn were seriously affected. The Speaker's private entrance to the Bundestag building could only be reached by boats. A mudslide on railway tracks near Grosskonigsdorf derailed the Ostend–Vienna express, killing six people.

In Britain, Noah's shadow lengthened as that oft-times soggy nation neared the Biblical 40 days and nights of rain. More precipitation had fallen on Norfolk during April through to June than since 1930. In the ensuing 62 days from March 14th, the London Weather Centre recorded only six days free from rain. Indeed, since the British government – through various agencies – began keeping weather statistics 256 years ago, only five springs have been wetter than that of 1983.

And the reason? The Meteorological Office in Bracknell, Berkshire, with its brand-new £6 million Cyber computer, carrying out 400 million calculations a second, had suggested that a ridge of high pressure was stuck over Siberia, with another over the Western Atlantic. They formed what one Met man described as a bowling alley, with rainy weather rolling down it to score endless 'strikes' on Britain.

Other Met men put the blame on stationary jet streams, as they had done in the past, which had caused the weather to lock itself into an unchanging pattern for weeks. This meant that incessant rain *or* prolonged heat could arise, and as 1983 moved into high summer it was clear that the previous extraordinary 'blocking high' of 1976 was repeating itself. Then, a drought in Europe had lasted 16 months, after which it continued to rain incessantly so that September–October 1976 became the second wettest period for that time of year since records began.

What causes a blocking high? The narrow, fast-moving bodies of air, about eight miles high in the upper atmosphere, are flanked by broader, slow-moving currents of air. They are formed by an exchange of energy between warm tropical and cold polar air. But of late, instead of these jet streams – sometimes known as circumpolar vortexes – moving slowly from area to area as usual, they have instead locked themselves into position over one geographical area.

Scientists are still not certain why they occur or what makes them suddenly break down. Computer models of the climate suggest that blocks are sustained by the outlying low pressure systems around both sides of the affected area which have the effect of pumping energy into the anticyclone, and this helps to keep it going.

Nevertheless it is clear they have played their own idiosyncratic role in upsetting the world's weather. In January 1978 America was the recipient of blocked circumpolar vortexes when a year-long drought came to an end, and four weeks of continuous rain was dumped on California. In Boston, Massachusetts, 21 inches of snow fell in one blizzard in February 1978 to be followed a few weeks later with a further 27 inches deposited by 100 mph winds.

A blocking high is also said to have plagued the US in 1980 when a blistering heatwave took more than 1,300 lives. At the same time South Africa froze, while Eastern Europe was drenched by sub-tropical rains. Several prolonged and biting recent British winters were the product of the same sort of blocking system.

Hubert Lamb believes he has detected a centennial cycle in the weather. A distinct pattern of storms at sea and severe flooding was accompanied by the cold winters of the eighties of the last centuries (1680, 1780, 1880, and so on). The present eighties, he feels, might not escape the trend. In addition the present extreme weather might be caused by a change in the direction of the wind, again the product of the vortexes. The milder westerlies that blow into Britain have been pushed southwards by the more biting northerly winds which leave their normal tight circular flow around the polar regions, and zig-zag in a slacker fashion, thrusting weather systems further south than they would otherwise be. Reid Bryson, of the University of Wisconsin, is one specialist who attributes the disappearance of the rains from the Sahel as a result of this southward drift.

The twentieth century is then a climatic watershed, standing at the threshold of momentous change, 'teetering', as Gordon Rattray Taylor put it in his *Doomsday Book* (published in 1970, his alarm at the way the weather was heading has been fully justified). And yet the new blocking weather syndromes are a joker in the pack. While the world hovers on the brink of a warming or a cooling, the flood/drought interface that they cause seem to defy long-term theorizing.

The argument has arisen from earlier chapters that we are still recovering from the last ice age, at its peak a mere 8,000 or so years ago, when the northern ice sheets blanketed Europe as far south as London. Scotland is said to be still rising from the weight of massive ice sheets. Scientists talk of an 'inter-glacial'. But even this is riddled with climatic fluctuations – a warm spell in the Middle Ages, followed by a prolonged cold snap known as the Little Ice Age. There was another sharp rise in temperature from the end of the nineteenth century until about 1940. Since then temperatures fell a little to level out in the sixties, followed yet again by a slight warming.

Whither the Weather?

Is, then, the world still warming via the greenhouse effect, or is it cooling as it heads towards its next ice age?

Climatologists, unfortunately for those who like clear-cut answers, still cannot

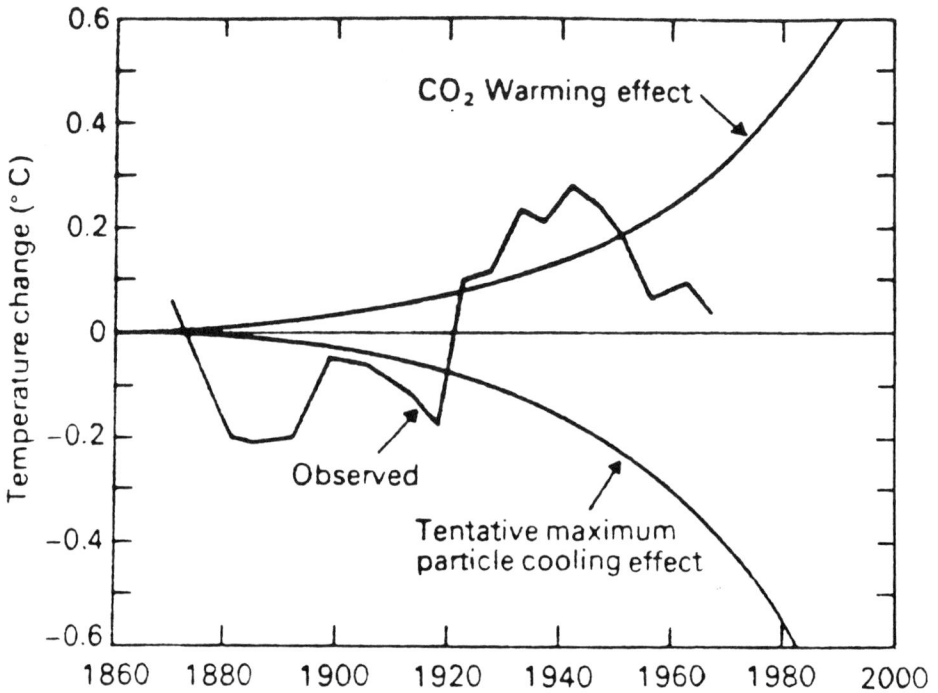

This diagram, by US climatologist Murray Mitchell, suggests that a cooling and a warming effect may cancel each other out.

agree. Reid Bryson is of the cooling school, and says that the world's cities are emitting as much particulate matter as a volcano. M.I. Budyko similarly points out that there has been a decrease in sunlight reaching earth of 4 per cent between 1936 and 1960, when smokestack industries were probably at their smokiest before lighter industries became more prevalent.

Others point to the evidence of observation. Since 1950 snow has been lying on the ground between 10 and 20 days a year, compared with only 6 days between 1910 and 1937. Mean winter northern temperatures are also down 0.8 °C from the previous half-century level.

But what about the greenhouse effect? Is the cooling cancelling out the warming? Quite possibly, as science writers are now in the comfortable position of being able to write books on either the warming or the cooling, or even to declare, as Lowell Ponte did, that the earth is both cooling and warming at the same time. Even the erstwhile and prolix John Gribbin had to defend himself from the charge of having it both ways when he published two books in one year (1983) arguing both for and against a warming. In his *Beyond the Jupiter Effect* he resolves the problem by arguing for a sequential effect: a cooling over the next two decades, followed by a warming in the next century.

But, whatever the trend, it seems the flooding problem is likely to get worse, and this chapter gives a broad hint at why this should be so. In short, a warming will increase precipitation and raise ocean levels, while a cooling will increase the risk of coastal storms and snowmelt floods. The storm-drought syndrome shows no sign of abating, thus prolonging all the short-range weather extremes, and even invert seasonal patterns with floods, heat-waves, droughts and cold snaps occurring at random. And all the while the climate teeters towards the twenty-first century, partly warming and partly cooling, until the real pattern at last reveals itself.

EPILOGUE

What is Being Done?

Seldom can flood risk alleviation be undertaken by isolated families, or even local communities. It needs government sponsorship and co-ordination.

Faced with the flood menace that seems to grow annually, many public authorities in the developed world have been obliged to take their responsibilities in this matter seriously. Since 1937, for example, the US government has implemented structural flood control and relief programmes at a cost of some $15,000 million. And the granting of aid on a large scale has led to an irreversible shift in public policy. It was a gesture towards the homeless and bereaved as well as a tacit promise that the government would henceforth have to do more than simply improve warning systems, or issue yet more floodplain prohibition notices to migrating homesteaders.

In other western countries, too, the authorities felt obliged to act in the aftermath of disaster. Much aid and restoration work was granted and initiated after the Lynmouth flood in England in 1952. The Arno floods of 1966 also provided the Italian authorities with the opportunity to undertake much needed schemes of urban renewal. A new sewerage system was installed and numerous buildings were rebuilt from behind crumbling but picturesque façades.

Governments can implement a variety of policies and approaches. First, however, conceptual and logistical problems need to be sorted out. Floods need to be *abated*. And abatement means identifying the causes of inundations and implementing the most obvious solution. For example, some public institutions with field experience can favour purely practical solutions such as halting and reversing deforestation or devegetation practices. The US Department of Agriculture prefers this approach.

Other attempts at flood control are focussed on effective management of the upstream watershed areas, either by preventing erosion or by diverting excess water into tributaries. Reforestation is an obvious example, as is the creation of catchment basins and reservoirs to trap sediment and debris. Measures to save New Orleans from flooding were taken early. In several pre-war floods levees were dug out several miles from the city to siphon much of the water down to the sea. At other times the floodwaters were diverted into Lake Pontchartrain.

Many authorities try to implement schemes of floodplain management. This is a regimen in which the main task is to keep a floodplain free of flood control structures wherever possible. Nature is left to its own devices, and artificial means of flood control are sparingly used. Flood control usually meant adopting the most effective method of controlling floods close to the place where they were most likely to do the most damage. The US Corps of Engineers favours this method.

In the past decade new requirements for the design and layout of buildings in floodplain areas, such as the Golden Triangle of Pittsburgh, have been implemented to reduce vulnerability. This had been done partly because construction companies have been deterred by the higher costs involved in meeting rigorous new building standards.

Buildings are nowadays constructed to obviate the worst excesses of floods. In some US cities tall buildings use the bottom few stories as open walled parking lots to allow the floodwaters to surge through. The planting of riverside vegetation helps. Contour ploughing is another option. This entails digging furrows along the slopes of elevations parallel with the track of the river which tends to hold any overflowing water that may otherwise run downhill. Bank raising, of course, is an obvious solution. Some can be in the form of permanent retaining walls. Others might merely be comprised of dumped rock, mattresses of brushwood or even emergency sandbags.

Not all of a river needs to be treated. The deep interlocking steel sheet piling that can be installed only on the sharp bends of rivers, rather than on the entire length, will enable the additional scour of the river bed to establish a new equilibrium. Other techniques involve the use of reservoirs, levees, channel improvements, flood spillways, pumping stations and floods forecasting.

Over the enormous riverine system in America various Federal agencies operate a variety of surveillance and forecasting techniques. Three major agencies, the National Bureau of Reclamation, the US Army Corps of Engineers and the National Weather Service, operate flood warning schemes. The National Weather Service co-ordinates the work of some 1,500 river-monitoring and 4,000 rainfall-measuring stations, many of them manned by private observers. Expert hydrologists at twelve key River Forecast Centres use computers to process dates, and prepare flood probability figures, and then transmit the information to the news media and emergency agencies.

Flood protection, however, is an expensive business; a sea wall can cost up to £1 million a kilometre. Expenditure in Britain since the disastrous East Coast floods of 1953 has been at least £80 million, and this includes the £7 million spent on the recently opened Thames Barrier. This expenditure is considered to be well worth it; there are savings from the damage to property and the loss of farming land, which may take up to five years to recover from prolonged sea water saturation. There is also the damage to public installations. Sewage works and electricity substations are particularly vulnerable, as are underground pipelines and cables.

For this reason, especially in the less developed countries, broadcast public warnings are preferred. Meteorological expertise is improving too; the remaining difficulties centering on locational factors rather than on timing or intensity. The Seismic Sea Wave Warning System (SSWWS) is beginning to produce results. Inaugurated after the 1946 Aleutian earthquake which sent tsunami shock waves through the Hawaiian Islands, it now covers the Pacific Ocean region. The system's big test came in 1964 after an Alaskan earthquake caused about $10 million worth of damage along the Californian coast, and flushed out 30 city blocks in Crescent City.

The great problem concerns tidal waves and cyclones. As we have seen in the case of Bangladesh in 1985, the tidal wave may overshoot the predicted area. If this happens enough times the authorities may be accused of crying wolf too often. Tsunami waves are particularly troublesome, since thay are often small and come in series. This puts the meteorologists in a dilemma. If they can predict the *precise* size of the wave, all is well. But people can't be expected to uproot and evacuate the area if the waves are too small to do any damage. On the other hand, once they have been pursuaded to move they must be dissuaded from returning too soon lest they be overwhelmed by the next wave in the series.

POSTSCRIPT

Throughout the summer of 1985 severe weather aberrations took their toll. Monsoon rains in June dumped more than 12 inches of rain on Manila in the Philippines' worst flood in more than a decade. The rains compounded the massive problems brought about earlier by Typhoon Hal, which had cut through the Philippines, leaving some 20,000 homeless and 50 people dead.

A few days later tidal waves on the south-west coast, and a landslide, killed 13 in Sri Lanka. A three-storey tenement building collapsed in Bombay during monsoon rains, killing at least 52 people and injuring 56. This particular building housed about 250 people, living as many as ten to a room.

Burst dams in China unleashed floodwaters that killed 47 and demolished several villages housing up to 30,000 dwellings. Crops over a wide area were devastated. For two whole months the torrential rains and typhoons battered the Far East. More than 500 died in floods in central China.

In early August damage from a flash flood which struck Cheyenne, Wyoming, during a thunderstorm, killed 12 people and wreaked damage to the value of $28 million. Cars and lorries were sent floating down streets under six feet of water. Telephone booths were torn out. A few days later it was the turn of the Soviet Union to suffer a similar fate. Heavy rains destroyed crops and inundated huge areas of farmland in the Khabarovks region of the Soviet Far East. Dozens of villages were cut off. The Soviet farming paper *Selskaya Zhizn* (Rural Life) said that in many areas five inches of rain had fallen in one day. In the first week of August rainfall almost reached the average for two months.

August continued to bring meteorological devastation to the United States as thousands fled from the path of Hurricane Danny as it swept up from the Gulf of Mexico with wind speeds of 80 mph, to later bring widespread rain and high tides. In the meantime two women were killed when the worst summer storms for a decade hit northern Germany. On the French Riviera a freak tidal wave chased 900 campers from sites near Marseilles. The River Inn overflowed at Innsbruck after a storm which raged through Austria, causing at least 10 deaths. All rail lines into the city were severed by landslips. The Danube was closed to shipping after doubling in depth, and many low-lying areas in eastern Austria were under water. Storms killed at least four people in Italy as floods, tornadoes and landslides afflicted northern and central parts.

BIBLIOGRAPHY

Allaby, Michael and Bunyard, Peter, *The Politics of Self-Sufficiency*, Oxford University Press, 1980

Allaby, M and Lovelock, J, *The Great Extinction*, Michael Joseph, 1983

Asimov, Isaac, *A Choice of Catastrophes*, Hutchinson, 1979

Asimov, Isaac, *From Heaven to Earth*, Dobson Books, 1968

Barnet, Richard J., *The Lean Years, Abacus*, 1974

Bellamy, H.S., *In the Beginning God*, Faber, 1945

Berlitz, Charles, *Mystery of Atlantis*, Panther, 1977

Berlitz, Charles, *Doomsday 1999*, Doubleday (USA), 1981

Berlitz, Charles, *Mysteries of Forgotten Worlds*, Corgi, 1974

Booth, Basil, and Fitch, Frank, *Earthshock*, Dent, 1980

Brooks, C.E.P., *Climate in Everyday Life*, Ernest Benn, 1950

Brubaker, Sterling, *To Live on Earth*, Mentor (USA), 1972

Bryson, R.A. and Murray, T.J., *Climates of Hunger*, University of Wisconsin (USA), 1977

Calder, N., *The Weather Machine*, BBC, 1974

Canning, John, *Great Disasters*, Octopus Books, 1976

Carr, Donald E., *Energy and the Earth Machine*, Abacus, 1978

Chorley, R.J., *Water, Earth and Man*, Methuen, 1969

Claiborne, Robert, *Climate, Man and History*, Angus and Robertson, 1970

Clube, Victor and Napier, Bill, *The Cosmic Serpent*, Faber, 1982

Cornell, James, *The Great International Disaster Book*, Charles Scribner Sons (USA), 1976

Critchfield, Howard J., *General Climatology*, Prentice-Hall (USA), 1974

Dacy, D.C. and Kunreuther, H., *Economics of Natural Disasters*, Collier-Macmillan (USA), 1969

Dasman, R.F., *Planet in Peril*, Penguin, 1972

Donelly, Ignatius, *Atlantis – the Antideluvian World*, Sidgwick & Jackson, 1970

Drake, W. Raymond, *Gods and Spacemen in the Ancient West*, Sphere, 1974

Ehrlich, Paul and Anne, *Extinction*, Gollancz, 1982

Engel, Leonard, *Sea*, Time-Life Books (USA), 1963

Goodavage, Joseph, *The Comet Kohoutek*, Pinnacle Books (USA), 1973

de Grazia, Alfred, *The Velikovsky Affair*, Abacus, 1978

Gregory, K.J. and Walling, D.E. (eds), *Man and Environmental Processes*, Dawson Westview Press, 1979

Gribbin, John, *Climatic Threat*, Fontana, 1978

Gribbin, John, *Our Changing Planet*, Abacus, 1979

Gribbin, John, *Genesis*, Dent, 1980

Gribbin, John, *Forecasts, Famines and Freezes*, Wildwood House, 1976
Gribbin, John, *Future Weather*, Pelican, 1983
Gribbin, John and Plagemann, Stephen, *Beyond the Jupiter Effect*, MacDonald, 1983
Halacy, D.S., *Ice or Fire?*, Barnes and Nobles (USA), 1980
Hare, R.M., *Restless Atmosphere*
Hassler, Gerd von, *Lost Survivors of the Deluge*, Signet (USA), 1978
Hawkes, Jacquetta, *History of Mankind*
Hewitt, R., *From Earthquake, Fire and Flood*, Allen and Unwin, 1957
Higgins, R., *The Seventh Enemy*, Pan Books, 1980
Hollis, G.E., 'Man's Impact on the Hydrological Cycle', *Geo Abstracts*, 1979
Holt and Langbein, *Floods* (USA), 1955
Hoyle, Fred, *Ice*, Hutchinson, 1981
Jensen, H.A.P., *Tidal Inundations Past and Present*, Nature Conservancy, 1953
Ladurie, Le Roy Emmanuel, *Times of Feast, Times of Famine*, Allen and Unwin, 1972
Lamb, Hubert, *Climate, History and the Modern World*, Methuen, 1982
Lamb, Hubert, *Climate: Present, Past and Future*, Methuen, Vol I, 1972, Vol II, 1978
Lamb, Robert, *World Without Trees*, Paddington Press, 1979
Lane, Frank W., *The Elements Rage*, David and Charles, 1966
Legget, Robert F., *Geology and Engineering*, McGraw-Hill (USA), 1962
Manley, G., *Climate and The British Scene*, Collins, 1952
Matthews, W.H., *Invitation to Geology*, David and Charles, 1971
McEvedy, Colin, *Penguin Atlas of Modern History*, Penguin, 1967
MIT, *Inadvertent Climate Modification*, MIT Press (USA), 1971
Mooney, R., *Colony: Earth*, Granada, 1976
Mooney, R., *Gods of Air and Darkness*, Souvenir Press, 1975
Moore, Patrick, *Can You Speak Venusian?*, Wyndham Publications, 1976
Moore, Patrick, *Countdown*, Michael Joseph, 1983
National Academy of Sciences, *Earth and Human Affairs*, (USA), 1972
Neurbergen, *Secrets of the Lost Races*, NEB, 1980
Olsen, Ralph E., *Geography of Water*, Wm C. Brown and Co (USA), 1970
Ponte, Lowell, *The Cooling*, Prentice-Hall, 1976
Orr, Walter and Lansford, Henry, *The Climate Mandate*, W.H. Freeman and Co (USA), 1979
Raikes, Robert, *Water, Weather and Prehistory*, John Baker, 1967
Sadil, Joseph, *Our Planet Earth*, Paul Hamlyn, 1968
Schneider, Stephen, *The Genesis Strategy*, Plenum Press (USA), 1976
Sears, Paul B., *Deserts on the March*, Routledge, 1949
Sears, Paul B., *Where There is Life*, Dell (USA), 1970
Shepherd, W., *Geophysics*, Rupert Hart-Davis, 1969
Smith, Keith, *Water in Britain*, Macmillan, 1972
Steiger, Brad, *Atlantis Rising*, Sphere, 1977
Steiger, Brad, *Mysteries of Time and Space*, Sphere, 1978
Stonely, Jack, *Tunguska*, Star Books, 1977
Taylor, Gordon Rattray, *The Doomsday Book*, Thames and Hudson, 1970
Tomas, Andrew, *Atlantis: From Legend to Discovery*, Sphere, 1973
Waltham, Tony, *Catastrophe*, Macmillan, 1978
Walworth, Frank, *Subdue the Earth*, Panther, 1977
Ward, Barbara, *Progress for a Small Planet*, Pelican, 1979
Ward, Roy, *Floods*, Macmillan, 1978
Warlow, Peter, *The Reversing Earth*, Dent, 1982

Warshofsky, Fred, *Doomsday*, Sphere, 1977

Wells, H.G., *A Short History of the World*, Pelican, 1982

Wendt, Herbert W., *The Romance of Water*, Dent, 1963

White, G.F., *Papers on Flood Problems*, Dept. of Geog., University of Chicago (USA), 1961

White, John, *Pole Shift*, W.H. Allen, 1980

Whittow, John, *Disasters*, Allen Lane, 1980

Wigley (ed), Ingram, Farmer, *Climate and History*, Cambridge University Press, 1981

Woolley, Leonard, *Excavations at Ur*, Benn, 1963

Velikovsky, Immanuel, *Earth in Upheaval*, Gollancz, 1956

Vogt, W., *Road to Survival*, Gollancz, 1949

Wijkman, Anders and Timberlake, Lloyd, *Natural Disasters – Acts of God or Acts of Man?*, Earthscan, 1984

JOURNALS and MAGAZINES

Yachting World, November 1982

Time Magazine, 28/3/83, 18/10/82, 7/2/93, 13/6/83, 3/10/83

Science, vol 219, January 1983

Astrophysical Journal, vol 248, October 1981

Icarus, vol 50, August 1982

Nature, vol 298, August 1982; vol 297, May 1982; vol 301, January 1983; vol 295, May 1982; vol 300, November 1982

Plain Truth, September, 1981

Sun Day (Sunday Express), 25/10/81

Birds, Spring, 1983

Readers Digest, August 1981

Sunday Times Magazine, 4/4/82

Geophysical Research Letters, vol 9, February 1982

PHOTOGRAPHS

ASSOCIATED PRESS: frontispiece, 19, 64, 72, 81, 100 (top), 104, 109, 110, 112, 126, 128, 129, 139

BBC HULTON PICTURE LIBRARY: front endpapers, 6, 11, 13, 17, 18, 23, 24, 28, 31, 47, 57, 62, 86, 92, 94, 95, 96, 98, 100 (bottom), 106, 116, 117, 120, 121, 125

MARY EVANS PICTURE LIBRARY: 137

POPPERFOTO: 21, 38, 51, 108, 113, 142, 148

PRESS ASSOCIATION: 59, 60, 85, back endpapers

INDEX

Adam 5
Adirondack Mts. 32, 113
Adriatic 63–65, 81
Aegean 45, 46
Afar, Ethiopia 77
Africa 12, 41, 43, 52–55, 64, 78, 136, 141, 145, 151
Agassiz, Louis 34, 35
Aguan valley, Honduras 73
Agung, Mt. 155
Ahuramazda (God) 10
Alabama 101
Alabama River 116
Alamagordo Creek, USA 77
Alaska 53, 67, 114, 132, 154, 163
Alexander the Great 49
Allaby, Michael 27
Allalin glacier 114
Allegheiny Mts. 109
Allegheiny River 109
Alps 80, 82, 127, 138
al'Ubaid, Iraq 48
Amazon, Amazonia 7, 45, 61, 140, 142–144, 146
America, Latin 10–14, 14, 25, 32, 34, 45, 53, 54, 73, 78, 95, 136, 141, 143, 145, 149, 151, 152, 154
America, United States of, 43, 48, 52, 53, 55, 61, 67, 71, 76–78, 93, 97, 101–118, 131, 132, 136, 138, 145, 146, 149–155, 157, 161, 162, 165
American Nat. Sc. Foundation 34
American Red Cross 110
Amerindians 7, 9, 13
Ancasmarca, Andes 8, 9
Andes 8, 9, 75, 152
Antarctica 32, 35, 53, 55, 148–150
Antilla island 14
Apennines Mts. 82
Appalachian Mts. 102, 109
Araucanian Indians 11
archaeology 45
Arctic 70, 133
Arid Zone Research Inst. 45
Arizona State University 149
Ark, Noah's 5–11
Arkansas 101, 103
Arkansas River 102
Arkhangelsk, USSR 133
Armana letters 7
Armenia 46, 137
Army Corps of Engineers (US) 101, 102, 108, 118, 130, 162

Arno River, Italy 64, 80, 82, 84, 141, 161
Arnold, Maurice 97
Arrarat, Mt. 8
Aryan race 5, 12, 14
Ascension Island 41
Ashurbanipal 1, 10
Asia 43–45, 49, 53–55, 61, 75, 78, 93, 95, 132, 136, 139, 141, 143
Asia Minor 12, 14, 46
Asian Inst. of Tech. 134
assam 67, 78
asteroids 52
astronomy, astrophysics 1, 34, 36, 40
Atchafalaya basin, river 102, 103
Atlantic 13, 14, 32, 41, 42, 45, 52, 53, 55, 61, 62, 67, 75, 79, 111, 112, 147, 156
Atlantic Oceanographic & Met. Labs 54
Atlantis island 12–15, 23, 40, 42, 43
Atlas Mts. 12
atmosphere 20, 25, 28, 35, 44, 50, 53, 74, 143, 148, 156
Attica, Greece 137
Australia 53, 54, 61, 67, 78, 151, 152, 154
Australian Nat. Univ. 35
Austria 127, 165
Austro-Hungary 92
Az 12
Azores islands 14, 41, 42, 45
Aztecs 8, 10, 12
Aztlán 12, 13

Babylon, Babylonia 5, 7, 20, 45, 48, 56, 123, 137
Bahamas 111
Bahamas Shelf 45
Bajo Uguan, Honduras 73
Baldwin Hills, USA 125
Baltic 80, 151
Bangkok 134, 135, 149
Bangladesh 1, 2, 69, 74, 95, 163
Barisal, Bengal 69
Barnet, Richard 46
Basque region 153
Basrah, Iraq 48
Bataks, Sumatra 10
Baton Rouge, USA 102
Bavarian Alps 46
Beaujolais district 156
Bellamy, Hans 19, 23, 40
Bendandi, Raffaele 68
Bengal 65, 58–70, 72, 73, 141, 149
Benicasa map 14
Berbers 12

Bergelmir (Noah) 11
Bering Sea 53
Bering Straits 54, 132
Bermunda islands 42
Bhola island, Bengal 68
Bible, The 5, 7, 15, 30, 123, 137
Biblioteca Nazionale, Florence 82
'Big Muddy', see Mississippi River
Big Sur, USA 154
Big Thompson canyon, USA 77
biosphere 50, 53, 144
Black Death 87
Black Sea 46
'blocking highs' 156, 157
Bochica (Noah) 8
Bolivia 152
Bombaby 145, 165
Bonn, Germany 93, 156
Bonnet, Charles 16
Booth, Basil 149
Borisov, P.W. 132
Bornholm island, Sweden 151
Borr 11
Boston, USA 76, 157
Brahmaputra delta 67, 68
Brandenberger, Arthur 36
Brassanone, Italy 81
Brazil 49, 52, 55–57, 59, 61, 63, 75, 79, 91, 97, 99, 113, 114, 130, 136, 141, 155–157, 162
British Columbia, Canada 113
Brookes, C.E.P. 44
Bronze Age 44
Brown, Hugh Auchincloss 25
Bristol Channel 61
Bryan, Richard 131
Bryson, Reid 43, 157, 158
Buckland, William 19
Budyko, Mikhail 148, 158
Buffalo Island, USA 102
Burghead Bay, Scotland 57
Bullard, Sir Edward 40
Bundestag, Germany 156
Buonarroti House, Florence 82
Bureau of Outdoor Recreation (US) 97
Bureau of Reclamation (US) 313, 162
Burma, 68

caesars 44
Cairo, Egypt 76
Cairo, USA 103
Calcutta 69, 70

Calder, Nigel 76
California 67, 93, 100, 101, 115, 117, 124, 131, 132, 152–154, 157, 163
California, Gulf of 131
California Junct. 113
Cambodia 44
Cambrian era 53
Cambridge, Ohio 77
Cambridge University, 22, 25
Cameron, Richard 34
Canaan 137
Canada 7, 32, 54, 71, 101, 102, 111, 113, 149
canals 134
Canvey Island, Essex 59, 60
Caputo, Michael 65
Caracerahy Rapid, Brazil 142
carbon dioxide 35, 75, 146–148, 150, 154
carboniferous era 54
Cardina 111
Caribbean 45, 71, 72, 105, 111
Carolina, S, USA 150
Carter, Howard 48
Carthaginia 13, 45
Caspian Sea 46, 130, 133
Castleton, Yorks. 78
catastrophism 15, 21, 30, 32, 33, 35
Cathedral museum, Florence 82
Caucasus Mts. 46
Cenozoic era 40
Chaldeans 10, 14, 48
Chandler wobble 25, 36
Chao Phraya River, Thailand 134
Charleston, S. Carolina 150
Chekian, China 87
Chengchou, China 90, 91
Cheyenne, Wyoming 165
Chiang Kai-Shek 91
Chibcha Indians 8
Chile 11, 66, 67
Chimborazo, Ecuador 152
China, Chinese 3, 7, 10, 11, 13, 16, 25, 43, 45, 46, 61, 66, 84, 87–97, 117, 138, 139, 149, 152, 153, 165
Chittagong, Bengal 68, 69
Cho-chia-kow River, China 91
Cholua (pyramid) 8
Chomoma, Honduras 72
Christians 11
Chumgmou, China 90
Chundu, Ecuador 152
Chungking, China 88
Chusien Chen, China 91
Cimbrian race 57
cities 145, 146, 148, 149
Clacton, England 56
Clairborne, Robert 37
climate, climatic change 25, 28, 34, 35, 38, 41, 43, 44, 49, 53, 132, 133, 136, 140, 147–150, 153–159
Climate Research Unit, England 155
Clube, Victor 27
Codex chimal-popoca 8
Colhuacan Mt. 8
Cologne, Germany 156
Colombia 8
Colorado, USA 45, 77, 130
Colorada River 102, 131

Colorado University 2, 35
Columbia University, USA 42, 54
Columbus 14
comets 20–24, 27, 29, 37
Comilla, Bangladesh 74
Conception, Chile 11
Conemaugh, Pennsylvania 122
Conemaugh Lake 119
Conemaugh River 119
Congress, Library of, USA 14
Connecticut River 111
Conquistadors, Spanish 8
Conway Valley, Wales 122
Cornell, James 105 115
Cornell University, USA 34
Cornwall, England 78
Corral, Chile 67
cosmos, space, etc. 15, 20, 22, 26, 34, 37, 38, 50, 61, 68
Costa Blanca, Spain 153
Crescent City, USA, 67, 163
Cretaceous era 55
Crete 42
critias, the 12
Cruz Laguna, Honduras 73
Curvier, Baron Georges 16, 53

Dakota, N, USA 32, 129
Dakota, S, USA 129
Dales, George F. 46
dams 88, 123, 124, 125, 129, 130, 132–134; Aswan High 130; Dale Dyke, Yorks. 123, 124, 130; Dickey Lincoln 131; Dolgarrog, Wales 122; Gezhouba, China 88; Hoover 131, 132; Koussou 130; Puentas, Texas 124; Rapid Creek, Dakota 129; Rybinsk, USSR 133; Sadd-el-Kajara 124; South Fork (Johnstown) 119, 120; Stava, Italy 127, 129; St Francis, Calif. 124; Tellicoe, New England 131; Teton, USA 127; Vaiont, Italy 126–128
Danes, Denmark 58
Danube River 45, 95, 137, 165
Dar-i-Khazinch 48
Darwin 15
Dayton, Ohio 110
Dead Sea 45
Debata (God) 10
deforestation 136–143, 146
Delaware Indians, 26
Delaware River 105
Delhi, India 155
deluge (flood) legends 5–15, 19, 27, 30, 33, 36, 41, 43, 46, 48, 49, 65
Dept. of Agric. (US) 161
deserts 44
Des Moines River 102, 105
Devon, England 78, 84, 85, 87
Dietz, Robert 54
dinosaurs 25, 29
Dolomites, Mts. 80, 82
Don River 132
Donelly, Ignatious 14, 40
Donting Lake, China 138
Dover, Straits of 57, 63
drift, continental 30, 40, 53, 54
drought 2

Dunn, William 54
Dunwich, Suffolk 58
Dutch, Holland 63
Dvina, USSR 133
Dwyfach 10
Dwyfan 10

Earth 1, 15, 20, 23, 25, 27, 30, 34–38, 50–56, 58, 61, 146, 155; tilt, pole shift, etc. 26, 27, 34, 36
earthquakes 1, 2, 9, 19, 23, 25, 32, 41, 43, 65–67, 127, 162, 163
East Indies 7
East Lyn River, Devon 84, 86
Easter Island 41
eclipses 23
ecology, ecosystem 25, 36, 44, 136, 140
Ecuador 151, 152
Egypt, Egyptians 12, 14, 16, 32, 44, 124
Ehrlich, Anne 131
Ehrlich, Paul 131
El Chichon (volc.) 77, 154, 155
El Nino 154
El Salvador 152
Emiliani, Cesare 35
Empire State Building 149
England 33, 44, 56, 57, 61, 63, 84, 99, 138, 143, 153, 161
English Channel 42, 55, 57, 67
Enlightenment 19
Environmental Protection Agency (USA) 3, 150
Eskimos 16, 23
Essex 49, 56, 59
Ethiopia 2, 5, 25
Euphrates River 5, 7, 45, 46, 48, 137
Europe 27, 32, 43–45, 49, 52–58, 75, 76, 78, 79, 80, 82, 93, 97, 114, 133, 136, 137, 138, 141, 149, 151, 153, 155–157
Everest, Mt. 53
evolution 27
Ewing, Maurice 54
Exmoor, Devon 84
Exodus 19

Faena, Italy 68
Fitch, Frank 149
Flemish Bight 44, 63, 149
Flohn, Hermann 75
flood control plans 110, 161; *see also* epilogue
flood legends *see* deluge legends
Florence, Italy 64, 80, 81, 141
Florida, USA 149
foraminifera 35
forests 44, 49, 95, 136–138, 140–46; *see also* deforestation
Fortunate Islands 42
fossils 16, 42, 146
Fort Wayne, USA 154
France, French 35, 55, 57, 63, 80, 93, 96, 108, 153, 165
Franklin, Pennsylvania 122
Frazer, Sir James 5

Gabinello Vieusseux, Florence 82
Galunggung, Mt. 155
Galveston, USA 71

Ganges delta 68, 97
Ganges River 69
Garonne River 156
Gautier Mills, Johnstown, USA 122
Genesis 5, 8, 10, 11, 15, 48
Geological Survey 117
geology, geophysics 15, 19, 25, 26, 30, 34, 36, 40, 46, 53, 55, 56, 83, 95, 123, 125, 127, 130
Geophysical lab. Faena 68
Georgia, USA 101
geosphere 140
Germany 44, 58, 93, 165
Gibraltar, Straits of 44
Gilgamesh, Epic of 5
glaciers 30, 32, 33, 35, 45, 46, 114
Glen Comfort, USA 77
Gold, Thomas 34
Gondwana 53, 54
Goodavage, George 21
Good Hope, Cape 67
Gough Islands 41
Grand Canyon, USA 12, 132
Great Lakes, N America 71, 153
Greeks, Hellens 5, 11, 12, 14–16, 43
Greenhouse Effect 3, 146, 150, 151, 158
Greenland 25, 32, 148
Gribbin, John 28, 155, 158
Grosetto, Italy 82
Grosskonigsdorf, Germany 156
Great Yellow River 89
Guangdon Province, China 88, 153
Guarani Indians 9
Guatemala City 1
Guayaquil, Ecuador 152
Gujurat, India 152
Gulf of Chihli, China 89
Gulf of Po Hai, China 87
Gunti River, Bangladesh 74

Haeckel, Ernst 40
Hai River, China 91
Haiti 111, 141
Halley, Edmund 22–24
Hampshire, England 56
Hapgood, Charles 25, 26, 35, 40
Hapi, flood god 43
Harappan culture 46
Hare, R.H. 2
Harris papyrus 16
Hatfield, Warren 132
Hatiya island, Bengal 68
Hawaii 9, 10, 66, 78, 162
Hawkes, Jacquetta 38
Hazard, Kentucky 109
Hebrews 5, 10
helium 20, 22
Heppner, Oregon 77
Hermitage papyrus 16
Herodotus 16
Hesperides island 42
Hewitt, Ronald 92
Hilo, Hawaii 67
Himalayas 67, 141
Hindus 10, 13, 16
Hiskia, King 123
Hittites 43
Hoerbiger, Hans 22

Hoffman, John 150
Hokkaido, Japan 67
Holinshead 58
Holland, Dutch 56–59, 149
Hollin, John 35
Holt & Langbein 105, 117
Homer 43, 56
Honan Prov., China 90, 95, 152
Honduras 72, 73, 95
Honshu, Japan 67
Hooghly River, India 69
Hopi Indians 8, 26
House of Commons, England 149
Hoyle, Sir Fred. 28, 38
Hrimthursar (God) 11
Hubei Plain, China 88
Hubei Prov., China 152, 153
Hudson River 55
Humber River 111
Hungarian Plains 44
Hungary 98
Huns 46
Hunter Valley, Australia 97
Huntingdon, Ellsworthy 43
Hurakan (God) 8
hurricanes, *see* storms
Hutton, James 30, 31
Hwang Ho (Yellow) River 84, 88–93, 97, 103, 138
hydrocarbons 20, 146
hydrogen 20, 22, 35, 147
hydrosphere 2, 51–53, 140

ice, ice ages 23, 25, 27, 29, 30, 32, 36, 36–42, 46, 50, 51, 53–57, 60, 83, 113–115, 134, 149–151, 157
ice caps, polar regions 3, 20, 25, 26, 30, 32–37, 41, 44, 54, 133, 147–150, 157
Iceland 20, 25
Idaho, USA 127
Illinois 97, 101–103
Illinois River 102
India 2, 7, 20, 45, 53–55, 67, 69, 70, 75, 76, 78, 97, 141, 151–155.
Indian Ocean 41, 52, 57, 78
Indiana, USA 153
Indians (American) 102
Indonesia 54
Indus civilization 43
Indus River 46
Indus Valley 45, 140
Inn River, Germany 165
Innsbruck, Austria 165
Inst. of Engineers (UK) 130
Inst. of Hydrology (UK) 99, 143
Ionian (Greek) school 15, 16
ions, ionizing 35
Iowa 9, 113, 140
Ipuwer papyrus 16
Iran 10, 14, 46
Iraq 45, 47, 48
Iraw Petrol Co. 48
Irish 16
Isahaya, Japan 78
Islands of the Blest 42
Israelites 7, 24
Italy 63, 65, 80, 125–127, 129, 153, 161, 165

Jamiolkowski, Michele 127
Japan 66, 67, 91, 132, 141, 143, 152, 153
Jefferson City, USA 110, 111
Jericho 45
Jerusalem 45, 123
Jiangxi Prov., China 152
Jibaro Indians 7
Johnstown USA 119, 122
Jordan River 124
Joseph, Emperor Franz 92
Joshua 124
Junagadh district, India 152
Jupiter 20, 37
Jurassic era 52, 55

Kaifeng, China 90, 91
Kamaish, Japan 66
Kane (God) 9, 10
Kano, Nigeria 78
Kansas 106, 110, 111
Kansas City 105, 111
Kansas River 105, 106, 110, 115, 117
Karachi 70
Kazakstan, USSR 132
Kazmann, Raphael 118
Keene College, USA 25
Kellog, William 148
Kent, England 63
Kentucky River 109
Khabarovks regions, USSR 165
Khan, Ayub 70
Khan, General Azay 70
Khasi Hills, Assam 78
Kish 48
Kitt Peak Nat. Observ. 37
Koblenz, Germany 156
Kolbe, P.W. 42
Krakatoa, Java 42, 67, 155
Kunreather, H. 115

Lacy, D.C. 115
Ladurie, Roy 114
Lamb, Hubert 44, 87, 155, 157
Lanchow, China 97
Lancing, Michigan 100
Lane, Frank 115
Laplace, Marquis de 22
La Reunion, Indian Ocean 78
Lebanon 137
Leggett, Robert F. 124, 125
Leibnitz, Gottfried 16, 17, 19
Lemuria, 40, 43
Lena River, USSR 114
Leningrad 149
levees 46, 83, 84, 89, 102, 103
Liaoning Prov., China 153
Library of Jewish Synagogue 82
Libya 12
Lili-nu-u (Noah's wife) 9, 10
Lincolnshire, England 56, 63
Lisbon, Portugal 66
Livingston, Wm. C. 37
Loire River 156
London 45, 57, 61, 73, 76, 78, 93, 99, 114, 145, 147, 149
Longarone, Italy 127, 128
Los Angeles, USA 125, 145, 154
Lossiemouth, Scotland 57

Lothair, Kentucky 109
Louisiana 61, 101, 103
Louisiana State University 102, 118

Maas delta, Holland 63
Mablethorpe, Lincs. 60
Macedonia 44
McCrea W. H. 37
McKinley, President 71
Madras 68
magnetic field 25, 28, 35
Mahabharata 13
Maine, Illinois 111, 130
Malagisma 13
Malaya 139
Malaysia 143
Mali civilization 44
mammoths 33
Manila 165
Manpura island, Bengal 68
Mao Tse-tung 138
Marajo island, Amazon 144
Mars 50
Marseilles, France 165
Massachussetts, USA 111, 143, 157
Massachussetts Inst. of Tech. 147
Mayas 12, 13, 16, 20
Medici chapel, Florence 82
Mediterranean 44, 45, 55, 64, 65, 80
Meghna estuary, Bengal 68
Meghna River, India 69
Melbourne, Australia 93
Memphis 45
Menes, King 124
Merak, Java 67
Merano, Italy 81
Mercury 50
Mesopotamia 5, 45, 48, 137
Mesozoic era 55
Messina, Italy 65
Meteorological Office (UK) 99, 156
meteors, asteroids etc. 23, 25, 27, 29
Mexicans, Mexico 8, 9, 13, 32, 44, 71, 79, 131, 132, 145
Mexico, Gulf of 35, 48, 61, 71, 77, 102, 108, 111, 115, 165
Miami, Florida 54
Miami River 110
Miami University 35, 42
Michigan 100, 153
Mid-Atlantic Ridge 40–42, 45, 55
Middle East, Levant 7, 10, 45, 55, 137, 141, 149
Midlands, England 131
Milankovitch, Milutin 37
Mineral Point, Pennsylvania 122
Minoa island 65
Minos, King 42
Minnesota 102, 105
Mississippi, USA 45, 61, 84, 102, 105, 108, 140
Mississippi River 90, 101, 103, 113, 114, 117, 118, 149, 153
Missouri, USA 101, 103, 115
Missouri River 102, 105, 110, 111, 113–117, 140
Mitchell, Murray 158
Molalla, Oregon 107

Mongolian 44
Monmouthshire, England 62
Mont Blanc 114
Montpelier, Vermont 103
moon 16, 61
Moore, Patrick 22, 40
Moray Firth, Scotland 57
Morgan City, USA 102
Moscow University 19
Moselle River 156
Mu, island 13, 43
Murray, J.J. 43
Muskingham River, Ohio 110
Mycenea 43

Nagasaki, Japan 78, 153
Nana-Bush (Noah) 26
Napier, Bill 27
NASA 150, 155
Nasir, Mt. 7
Nat. Academy of Sc. (US) 150
Nat. Climatic & Atmospheric Research Inst. (US) 36. 48
Nat. Inst. of Amazonia Research 144
Nat. Oceanic & Atmospheric Inst. 154
Nat. Water Council (UK) 99
Nat. Weather Service (US) 162
Navajo Indians 12
Nazis 22
Nebraska 130
Neckar River, Germany 156
Negev Desert 45
Neolithic era 33, 56
Neosho River, USA 110
Neptune 50
New England 32, 111, 131, 138
Newfoundland 71
New Guinea 35, 43
New Jersey 153
New Mexico 77
New Orleans 61, 102, 105, 108, 118, 155, 161
New South Wales, Australia 97
New York City 145, 150
New York State 76, 103, 111, 113, 115
New Zealand 67, 138
Nile delta 45, 46 48
Ninevah 45
Nippur 7
Noah 5, 6, 13, 15, 19, 30, 48
Noakhali, Bengal 69
Nordic tribes 46, 55
Norfolk, England 56, 156
Normans 58
North Sea 56, 57, 59, 63, 79, 80, 153
Norway 57
Nova Scotia 111
Nutalaya, Prinya 134
Nu-Wah (Noah) 10

Ob River, USSR 114, 133
Oceanus (God) 56
OECD 146
Ohio 153
Ohio River 102, 103, 109, 110, 117
Ohio State University 36
Oklahoma 9, 101, 140
Old River connection, USA 102, 108

Omar mosque 123
Onega basin 133
Onega, Lake, USSR 133
Ontario 111
Ontario, Lake 111
Oregon, USA 107
Ovid 16
oxygen 35

Pacific Coast Highway, 152
Pacific Ocean 13, 32, 41, 52, 61, 66, 67, 71, 75, 132, 152, 154, 162
Pakistan 45, 69, 70, 75
Pakistan Met. Dept. 73
Paleocene era 55
Palermo, Italy 153
Panama 67
Pangea 40, 53, 54
Panthalasia 53
Paraguay 151
Paris 16, 44, 93, 96
Pavia, Italy 125
Pechara basin 133
Peirium, King 13
Peking, China 88
Pellestrina island 64
Pennsylvania 105, 112, 115, 119, 153
Perevedentsere, Vladimir 133
Permian ice era 54
Persian Gulf 45, 48, 137
Persopolis 45
Peru 7, 9, 75, 151, 152
Phaeton 16
Pharoes 137
Philippines 136, 143, 145, 165
Phoenicians 14, 137
Phrygian race 44
Piave River 126, 127
Piazzade Duono, Florence 81
Piccadilly, London 149
Pico island 42
Pillars of Hercules 12
Pitdaea, Greece 65
Pittsburgh, USA 162
Plagemann, Stephen 28
planetsimals 22, 27, 28, 52
Plato 12, 16, 18, 137
Platte River 130
Po River 45, 63, 81, 149
pole shift theories, *see* Earth tilt
pollution 147
Polynesia 7, 151
Pontchartrain Lake 161
Ponte, Lowell 44, 132, 158
Ponte Vecchio, Florence 80–82
population 145
popul vuh 8
Port Jervis 108
Portugal 55
Prestwich, Joseph 33
Priene, Asia Minor 124
Puira, Peru 152

Qingjun, Dr H. 138
quaternary era 54, 55
Quebec, 111
Quetzalcoatl 13

Index

Quiche Mayas 8
Quito, Ecuador 152

Raikes, Robert 93
rainfall, precipitation 31, 35, 37, 42, 44, 48, 49, 51, 52, 74–78, 82, 83, 95, 105, 109, 110, 119, 122, 123, 130, 140, 141, 143, 144, 152, 154, 156, 159, 165
Rapid City, Dakota 77, 132
Reagan, President 132
Red Cedar River 100
Red River 102
Reservoirs Acts (UK) 130
Revelation, Book of 9
Rhine River 45, 82, 156
Rhone River 156
Rig-Veda 10
Riks museum, Stockholm 42
Rio de Janeiro 151
Rio Grande de Sul, Brazil 151
Rio Grande River 102
River Forecast Center (US) 162
rivers 25, 41, 45, 46, 48, 52, 56, 71, 80–99, 101–118, 123, 124, 134, 138, 153
Roanoke River 117
Rock Island, Illinois 97
Rocky Mts. 16, 102, 131
Romans 16, 44, 45, 56, 93, 123, 138
Rome University, 65
Ross ice shelf 149
Royal Society 19
Russia 46, 91, 114, 130–133 145, 149, 165
Russian Academy of Science 42

Saar River 156
Saas valley 114
Sac and Fox Indians 9
Sacramento River 100
Sadil, Josef 52
Sagan, Carl 20
Sahara desert 75, 141
Sahel 2, 157
St Helens, Mt. 155
St Louis, USA 102
St Marks Sq. Venice 64
Salati, Eneas 144
San Andreas fault 101, 125
San Croce, Florence 81
San Diego 132, 154
San Firenzi palace, Florence 82
San Francisco, USA 66, 67
San Frediano, Florence 81
San Pedro Sula, Honduras 73
Sanmen Gorge, China 95
Sanriku Beach, Japan 66
Santa Caterini, Brazil 151
Santa Croce, Florence 82
Sargasso Basin 55
Saturn 50
Scandinavia 11, 32, 49
Scheldt delta, Holland 63
Schowen, Holland 63
Schneider, Stephen 36, 145
Scioto River, Ohio 110
Scotch cap lighthouse 66
Scotland 57, 63, 138, 157

Scotsville, USA 77
Scranton, Pennsylvania 112
Scripps Oceanic Inst. 147
seas, oceans 25–29, 33–37, 41, 45, 52–61, 65, 75, 84, 135, 140, 144, 147, 149, 154, 159, 162
Seine River 45, 82, 95, 96, 138, 156
Seismic Sea Wave Warning System (SSWWS) 162
Severn Basin 61, 62
Shaanxi Prov., China 88, 153
Shang City, China 88
Shang Dynasty 88, 138
Shanghai, China 88, 138
Shantung Prov., China 89
Shrewsbury, England 97
Siberia 52, 55, 114, 133, 156
Sicily 78, 153
Sichuan Prov., China 88, 138
Sichuan River 152
Sigebert, King 58
Simpson, Sir George 37
Singapore 76, 78
Sinton, Texas 104
Skaw, The 80
Smith, Norman 124
Snake River, Idaho 127
Snowdonia, Wales 122
Solomon 123
Solon 12
South Africa 151, 157
South Seas 13
Spain, Spaniards 7, 8, 11, 13, 55, 77, 93, 141, 153
Sri Lanka 165
Stanford University 131
Stava, Italy 127
Storms, cyclones 3, 49, 52, 58, 64, 68–77, 80, 95, 101, 104, 111, 151–154, 157, 159, 165
Stuiver, Minze 147
Subboreal era 20, 57

Suffolk 58
Sumarians 5, 45, 46, 48, 56, 93, 137, 140
Sumatra 10, 67
sun, solar radiation 16, 37, 50, 61, 68, 70, 75, 146, 149, 158
Sunda Strait, Java 67
sunspots 38, 68
surge tides, see tides
Sussex University 37
Sweden 151
Switzerland 127
Sylhet, Bangladesh 74
Syria 7
Szeged, Hungary 92

Tamandere (Noah) 9
Tambori 155
Taviani, Sr. 81
Taylor, Gordon Rattray 141, 157
Tegucigalpa, Honduras 72
Teheran, Iran 77
Tennessee, USA 131
Tethis Sea 53, 55
Teutons 57
Texas 101, 105, 124, 153

Tezpi (Noah) 8
Tezxaltipoca (God) 8
Thailand 134, 136, 143, 153
Thames Barrier 57, 99, 162
Thames River 45, 56, 63, 83, 94, 114
Thebes 45
Thera island 42, 65
Third World 1, 73, 91, 95, 141
Tibet 87, 88
Tiehtse (Yangze) Kiang River, China 61, 87, 90, 91, 93, 138, 153
Tientsin, China 87, 89, 91
tides, tidal waves, tsunamis 9, 11, 28, 33, 40, 60–70, 150, 152, 163
Times, The 91
Tioza River, Italy 92
Toc, Mt. 127, 128
Tokyo, Japan 66, 67, 145
Toltecs 14
Tonga 153
Toon, Brian 150
Topeka, Kansas 110
Toronto 111
Toronto University 2, 34, 35
Trenton, Italy 81, 108
Triassic era 55
Tristan da Cunha 41
Troano manuscript 13
Troy 43
Tucson, Arizona 37
Tungkwan gorge, China 97
Tunguska, Siberia 27
Turkey, Turks 46, 137
Turia River, Italy 93
Turin University, Italy 127
Tuscarora Beach, Japan 66
Tsinam 89

Uffizi gallery, Florence 82
Ulna valley, Honduras 73
UNCTAD 143
Uniformitarians 30, 31, 34
Unionville, Maryland 77
Upanges 10
Upper James River, Virginia 77
Ur 47, 48
Uranus 50
Urichar island 74
Uruguay 151
US Office for Disaster Assistance 2
US Weather Bureau 77, 109, 115, 117
Utah, USA 113
Uttar Pradesh, Bangladesh 74
Uzbekistan 132

Valencia, Spain 93
Vancouver, Canada 113
Velikovski, Emmanuel 15, 19–21, 25, 27, 40
Venezia, Florence 81
Venezuelans 13
Venice, Italy 63–65, 80, 149
Venus 37, 50, 150
Vermont, USA, 103
Vikings 43, 57
Villa Floriano, Italy 127
Virginia, USA 77
Vogt, William 140

volcanoes 9, 19, 28, 34, 35, 41, 42, 54, 67, 79, 154, 155
Volga River 45, 55, 130, 132, 133

Waialiali, Mt. 78
Walcheren island 56
Wales 62, 79, 99, 122
Walworth, Frank 41
Wanthalia district 153
Warlow, Peter 33
Washington, USA 93, 131, 132, 151, 153
Wegener, Alfred 40
Wells, H.G. 44
West Java 67
West Lyn River 84, 86
West Tilbury 56

Weyer, Edward 34
Whewell, William 19
White, John 25
White Sea 133
Whittow, John 97
Willard, Utah 108
Willow Creek 77
Wisconsin University 157
Wisley, Thomas 46
Woodvale, Pennsylvania 122
Woolley, Leonard 47, 48
Woolwich, England 57
Wuhan, China 138
Wyoming 165

Yangste, *see* Tientse River

Yankeetown, Florida 76
Yazoo River 102
Yellow River 89–91, 102
Yellow Sea 89
Yemen 124
Yenisey (Yenisi) River 133
Yokohama Park, Japan 67
Yorkshire, England 123
Yucaton 44
Yugoslavia 63, 80, 127

Zapotecs 8
Zarate, Augustus de 9
Zeeland 63
Zuider Zee 58